Self-Sufficiency
Household
Cleaning

Rachelle Strauss

NEW
HOLLAND

First published in 2009 by New Holland Publishers (UK) Ltd
London • Cape Town • Sydney • Auckland

Garfield House	80 McKenzie Street	Unit 1	218 Lake Road
86–88 Edgware Rd	Cape Town 8001	66 Gibbes Street	Northcote
London W2 2EA	South Africa	Chatswood	Auckland
United Kingdom		NSW 2067	New Zealand
		Australia	

ISBN 978 1 84773 458 7

Senior Editor   Corinne Masciocchi
Design   e-Digital Design
Main illustrations   Michael Stones
All other artwork   e-Digital Design
Production   Laurence Poos
Editorial Direction   Rosemary Wilkinson

3 5 7 9 10 8 6 4 2

Reproduction by Pica Digital PTE Ltd, Singapore
Printed and bound in China by Toppan Leefung Printing Limited

Disclaimer

The author and publishers have made every effort to ensure that all information
given in this book is safe and accurate, but they cannot accept liability for any
resulting injury or loss or damage to either property or person, whether direct
or consequential or however arising.

CONTENTS

Foreword

What a refreshing book – and a great way to release yourself from the need to spend huge amounts on cleaning products. This book is like having a friend recommend the best products to clean your home with in a healthy, safe and economical way, and work towards self-sufficiency. Without getting too scientific, you'll find simple explanations of why certain hazardous chemicals are so bad for us, and the great news is there are some excellent alternatives, from getting out the elbow grease and going back to tried-and-tested ingredients such as lemons and white vinegar, to buying products that are more eco-friendly and sustainable and don't cost the earth, in every sense of the meaning. You'll find clear, easy-to-follow cleaning 'recipes' to replace many commercial products, along with innovative suggestions for saving money and ticking the green box in the process. If you have young children, it's liberating not to need a child lock on the cleaning cupboard because everything is natural!

Over the years and throughout my mission to seek out natural alternatives to just about everything, I've seen many books that can sometimes give me 'green fatigue', but not this one. There's no preaching as to what must be done, rather a gentle persuasive nod towards making small changes which in turn will make a big difference.

This book is perfect for anyone dipping their toes into the 'eco' world and for those who care about the health and wellbeing of their family and the planet. It's also ideal for anyone on a budget, and what better way to save money than to ditch the chemicals from your shopping trolley and look first to what's already in your kitchen cupboards or growing in your garden.

You'll be amazed how some of the ideas featured here can change your life. You'll also probably feel better and suffer less fatigue, fewer headaches and skin conditions by making only a number of relatively minor changes. I hope you enjoy learning all about the green approach to cleaning!

Janey Lee Grace
Author of Imperfectly Natural
www.imperfectlynatural.com

Introduction

Welcome to the world of safer, greener
household cleaning. Cleaning that will help you
become more self-sufficient and benefit your health,
the environment and your bank balance.

In the beginning...

I started making my own cleaning products when I was pregnant
with my daughter. Being responsible for bringing new life into the world
led to several lifestyle changes because I wanted to give my baby the best
start in life. I was determined to have an organic baby! I couldn't believe
how easy it was to make these products! They actually worked and saved
me money. If that wasn't enough, I was saving packaging from the landfill and
keeping toxic chemicals out of the environment. And my home looked and
smelled great, too.

Benefits to health

But there was another benefit. I'd had a number of health conditions, such
as headaches, frequent infections, stuffy nose, itchy eyes and sore throat for
years. I'd learned to live with them, thinking that most people felt the same
way. But within a few months of weaning myself off conventional cleaning
products, my health improved dramatically. All those minor ailments
I'd been living with had vanished, along with the air
pollution in my home. That's why I'm delighted to
be able to share with you some of the secrets
I've discovered about safer,
greener and more self-
sufficient household cleaning.

What you will learn from this book

This book will provide you with:
- Information about harmful ingredients used in some conventional household cleaning products.
- Details about using safer, readily available ingredients for cleaning.
- Simple, effective recipes for everyday cleaning jobs.
- Preventative cleaning methods so that you never have to do another spring clean!
- Ways to save money, space and resources.
- Tips to protect the environment.
- Information about ways to help improve your own and your family's health.

Keep it simple

There are many books available that teach you how to make your own cleaning products. This book is different in that it keeps things very basic and simple, using natural products that are commonly available. You'll find a product to suit every cleaning need, with the reassurance of knowing that it has been tried and tested for years!

Downsizing your product range

You probably think you need a huge range of products to clean your house: one to clean the floor, one to clean the bathtub, another for the kitchen sink, specialized ones for worktops, formulas for glass, different ones for polishing wooden furniture, and yet more for drains and ovens. But you really don't need all these different products. I use one simple household ingredient to clean worktops, sinks, the bathtub and hob, and it's so safe you can eat it!

Avoiding overwhelm

Do you remember the last time you bought a cleaning product and the store didn't stock your usual brand? Did you stand in front of a bewildering array of products and find you were overwhelmed with choice? Did you get your product home, start using it and find it didn't work in the way you expected? You just knew you should have bought the one with the citrus scent, or the blue label, or the trigger spray. This book will help take away the decision-making process with my tried and tested recipes. It will save you time and expensive mistakes.

Saving time, space and money

It's a competitive world out there! Manufacturers tempt us to buy products with the promise of something new: a new fragrance, new formula, new cleaning power, new packaging, new label… All this choice can lead to confusion, and you may end up buying a product that is not suitable for the job.

Once you've honed down your cleaning supplies, you'll discover that a few simple and effective ingredients are enough to keep your entire home clean, safe and smelling great. Most of the recipes featured here are far cheaper than shop-bought products because you are not paying for packaging and marketing. By the time you've finished reading this book, you'll have learned how to decipher product labels and you'll discover just what you've been paying for all these years. You'll never look back!

A small price to pay for huge rewards!

Homemade formulas are not as processed as shop-bought ones, so they may separate. If this is the case, don't worry; it doesn't mean they've gone off. Just shake or mix them before use. Equally, you might need to scrub a little harder or leave things to work, but you'll get enviable toned arms in the process and healthier air in your home! Make the most of the situation and make yourself a cup of tea or thumb through your favourite book whilst the products do the work.

Are you ready to discover ways to clean your home in a healthier, safer and more economical way? Then let's learn all about green cleaning.

Green cleaning

What is green cleaning? Green cleaning involves cleaning your home without harming the environment or your health. It is an empowering step to take if you want to be more self-sufficient. Green cleaning means moving away from conventional products containing synthetic chemicals, or petroleum-based cleaning agents, and using safer ingredients instead. Many store cupboard ingredients can be used for safe, natural and effective cleaning.

Prevention is better than cure

A major aspect of green cleaning is taking preventative measures. If you do not allow dirt, grease and germs to build up, you will not need harsh chemicals to deal with them. These simple but effective tips will save you hours of cleaning:

• Trap and remove dirt and pollutants before they enter your home with good-quality doormats.

• Regularly dust and vacuum to minimize particles in the air, which can lead to allergies.

• Wipe around kitchen and bathroom surfaces on a daily basis so that greasy spots and soap scum do not build up.

• Dry surfaces properly so that bacteria cannot multiply, reducing the risk of mould and mildew forming.

• Wash your hands before preparing food to reduce the need for antibacterial products around your home.

• Keep areas such as door handles, toilet flushes and light switches clean to give germs less chance to breed.

Conservation

Green cleaning takes into account conservation of water and energy too. For instance, only using the washing machine when you have a full load and using as cool a wash as possible will not only help the environment, it will save you money too. Most lightly soiled laundry can be washed at 30 degrees with very good results.

Why should I green clean?

There are many reasons to green clean.

Safer for you

Many cleaning products contain a bewildering array of synthetic chemicals. Some are safe, others are not. Immediate effects of some of these chemicals include stinging or watery eyes, sneezing fits, a choking sensation, headaches, asthma, skin irritation and nausea.

Ongoing research shows that long-term exposure to some of these chemicals has been linked to more serious health conditions, such as hormone disruption, various cancers, central nervous system disruption and debilitating 'modern' illnesses, such as multiple chemical sensitivity. Scientists now believe that indoor air pollution is two to five times worse than outdoor air pollution, due to the amount of chemicals we use in our homes. That's shocking isn't it? Many of us are living in a chemical soup we have brought into our homes, believing that these products are keeping our homes clean and safe!

Safer for your family

If you have children or pets, they will benefit from a green, clean environment. Animals and children breathe much faster than adults and have smaller bodies, so the impact of inhaling and ingesting chemicals is much stronger and potentially harmful. In addition, they spend a lot of time playing on the floor and putting things in their mouths that can have chemical residues on them.

Safer for the environment

If you make your own cleaning products, you will have less packaging to dispose of. Many shop-bought products come in plastic packaging which cannot be recycled and ends up in the landfill where it stays for hundreds of years. Tiny pieces of plastic have been found on the ocean floor where they can cause devastation to wildlife. Mistaken for food, plastic can become lodged in the stomachs and throats of marine life, resulting in death. Additionally, by making your own products, you'll know exactly what you are pouring into the water system. Every cleaning product we use eventually ends up in the environment at large.

Cheaper for you

You will save lots of money by making your own cleaning products! Most recipes can be made from basic household ingredients. Before the Second World War, housewives used baking soda, lemons and vinegar to do most of their cleaning.

Saves space

Finally, making your own cleaning products frees up cupboard space! You'll find that you only need a small range of ingredients to cover every cleaning job in your home – safely, efficiently and naturally.

As you go through the various cleaning jobs in this book, you'll realize that many products are all-purpose and you don't need a separate product for each task.

Basic rules

Making your own household cleaning products is fun and rewarding, but bear in mind that natural doesn't necessarily mean safe. Turpentine, used for thinning oil-based paints and producing varnishes, is a natural product obtained from the resin of trees. However, its vapour can burn skin and eyes. When inhaled it can damage the lungs and central nervous system, and if ingested, can cause renal failure. You'll be pleased to hear that many of the ingredients used in this book are safe to ingest, such as vinegar, bicarbonate of soda and salt. Yet even these substances, in large enough quantities, can be toxic! Follow these simple rules when making your own cleaners to ensure you and your family stay safe and healthy.

✗ Never mix different shop-bought household cleaners or solvents together. For example, bleach combined with household cleaners containing ammonia, such as toilet cleaners, produces a toxic and carcinogenic mix which can be fatal if breathed in.

✗ Never reuse shop-bought cleaning containers as, even if rinsed out, they may still contain harmful chemical residues that might react with your own homemade products.

✗ Never mix homemade cleaners and conventional products; you never know what fumes will be given off.

✓ When you make your own household cleaners, label the containers and list all the ingredients.

✓ Keep all household cleaning products out of reach of children and pets.

✓ Wear rubber gloves where possible; even bicarbonate of soda can dry out your hands!

Now you've remembered these simple precautions, we can learn about some of the toxic ingredients in some conventional household cleaners.

Toxic ingredients

Many chemicals are used in commercial household cleaning products. Some of these can exacerbate conditions such as asthma; others may affect the skin if they come into contact with your hands or get splashed into your face, causing burns, rashes, irritation or dryness. Some chemicals have an adverse effect on the environment by polluting waters and the air. Here are a few you may have heard of and may wish to avoid.

Chlorine

Which products contain chlorine?
Chlorine is found in some multi-purpose cleaners, liquid bleaches, toilet bowl cleaners, washing powders, disinfectants and mould inhibitors. It is a general biocide.

Can chlorine harm me?
Chlorine irritates the eyes and lungs, can trigger asthma attacks, and aggravate respiratory ailments or heart conditions. It dries the mucous membranes and can burn the skin. If you accidentally splash it on dark clothing when working, chlorine bleach can take out the colour in the fabric. The harmful effects are intensified when the fumes are heated, such as from chlorinated water in the bathroom when you run a bath or shower or when you open the dishwasher door and get a waft of steam that contains chlorine.

Does chlorine harm the environment?
The 1990 Clean Air Act lists chlorine as a hazardous air pollutant. It is argued that chlorine is a safe product in the environment as it breaks down into harmless salt and water. The real issue is that the by-products of chlorine (organochlorines and dioxin) do not break down readily and remain in the environment.

These products get dumped into our streams and waterways, causing polluted waters. As a result, the fish become contaminated, animals eat the fish and humans eat the animals, creating a concoction of harmful chemicals in the food chain. It is believed that many of us ingest a daily amount of dioxin 300 to 600 times greater than the Environmental Protection Agency's 'safe dose'. When consumed, these chemicals accumulate in the fatty tissues, causing hormonal imbalances, diabetes, cancer, suppressed immune systems, endometriosis, reproductive disorders and detrimental effects on foetal development in the womb.

Detergents

Which products contain detergents?

Detergents are the main component of many household cleaners. You will find them in heavy-duty cleaning products that deal with food, grease and oily residue, such as dishwashing and laundry products. Detergents contain other components such as surfactants, enzymes, acids, caustics, bleaching agents and optical brighteners. These components all have their part to play in dislodging dirt and grease from surfaces and fabrics around your home.

Can detergents harm me?

The human body contains many sensitive mucous membranes located in the eyes, nose, mouth and lungs. These require a delicate balance between water content and other bodily fluids such as sebum. If any of these mucous membranes comes into contact with detergents, this delicate balance can be disrupted. The moisture-retaining sebum is oily, and can be attacked by detergents because they are designed to dissolve oil. Once this happens, tissues can lose their natural elasticity and become easily torn or ruptured. You might notice that your skin is dry or irritated after using certain products, but for some people the outcome can be more serious and result in dermatitis or eczema.

It is easy to protect your hands, with gloves. However, many modern cleaning products come in spray bottles or require dilution in hot water. This leads to chemicals being sprayed in the air or being transmitted in hot, steamy fumes which can be inhaled into the lungs or reach the eyes and mucous membranes.

Formaldehyde

Which products contain formaldehyde?

Formaldehyde is used in a wide variety of products found in our homes. It's a chemical used as a binder and preservative in literally hundreds of household products, including paper products like toilet rolls, sanitary protection, tissues and building materials. Additionally, formaldehyde is added as a disinfectant to some cleaners as it kills most bacteria and fungi.

Can formaldehyde harm me?

Formaldehyde resin is used in many construction materials which makes it one of the more common indoor air pollutants. It has been shown to irritate the eyes, throat, skin and lungs, cause nausea, nosebleeds, insomnia, headaches, coughing, wheezing, and can trigger asthma attacks, lower immunity, and cause fatigue and skin rashes. Formaldehyde is classified as a probable human carcinogen. The International Agency for Research on Cancer (IARC) has determined that there is 'sufficient evidence' that occupational exposure to formaldehyde causes nasopharyngeal cancer in humans (the upper part of the throat, behind the nose). This type of cancer spreads widely and produces few symptoms early, which means that most cases are quite advanced when detected.

Does formaldehyde harm the environment?

Formaldehyde is one of the large family of chemical compounds called volatile organic compounds (VOCs) which you'll learn more about in the following chapters. Levels of formaldehyde emitted from products increase with temperature and humidity, so bear this in mind if you use a disinfectant or floor cleaner containing formaldehyde that you add to a bucket of hot water.

Fragrance

Which products contain fragrance?
Nearly every cleaning product is scented. Most fragrances or 'parfums' are made from a cocktail of petrochemicals. Several thousand different chemicals can make up one fragrance and manufacturers do not have to disclose what ingredients they use.

Can fragrances harm me?
If you react badly to the smell of a product, then listen to the warning your body is giving you! If you find yourself with itchy eyes or a tickly throat as you walk down the cleaning aisle of your supermarket, then maybe you'll begin to understand how some of these products are affecting your health. Inhalation of fragranced products can cause immediate symptoms such as sneezing, itchy eyes, headaches, nausea and wheezing. More chronic symptoms from exposure to artificial fragrances include asthma, multiple chemical sensitivity, lethargy and short-term memory loss.

How else can I be affected by fragrances?
Fragrances don't just enter your body through the nose. They are also absorbed through the skin. Two products that are notorious for causing skin irritation are laundry detergents and fabric conditioners. Both are highly scented and residues of the chemicals used to make the fragrance are against your skin all day in the clothes you wear.

Fragrances serve no useful purpose in a product in that they do not make them more effective. They are there to 'brand' a product and cover up some of the terrible smells of the raw ingredients. Fragrances are designed to linger – on your clothes, bedding and in the air of your home – so you are constantly in contact with them.

Triclosan

Which products contain triclosan?

Triclosan is one of the most widely used antibacterial and antifungal ingredients in household cleaners and personal care products. You will find it in some hand- or dish-washing soaps and even impregnated into surfaces such as food storage boxes and 'antibacterial' chopping boards.

Can triclosan harm me?

Once touted a 'cure all' for germs and bacteria everywhere, there is now concern about the dangerous consequences of triclosan when it is absorbed through the skin and accumulates in the body. Triclosan has been found accumulated in fish tissue and human breast milk. One of the most devastating side effects of the overuse of antibacterial products is that resistant strains of bacteria are developing all the time. This means that bacteria that were once killed by triclosan have found ways to mutate and survive in much the same way as some 'superbugs' are now resistant to antibiotics.

How else can I be affected by triclosan?

Research shows that when triclosan meets with free chlorine in tap water it produces a number of toxic products such as 2,4-dichlorophenol. In the presence of sunlight, these products can produce dioxins. Some dioxins are extremely toxic and are potent endocrine disruptors.

Reports have suggested that triclosan can combine with chlorine in tap water to form chloroform gas, which the United States Environmental Protection Agency classifies as a probable human carcinogen. Scientists are concerned that if you use a product containing triclosan, such as antibacterial soap, and then jump into a chlorinated swimming pool, it could produce dioxin on the surface of your skin that then gets absorbed into your body.

Volatile organic compounds (VOCs)

Which products contain VOCs?
Many liquid cleaners and household products contain organic solvents such as ethanol, methanol, isopropyl alcohol, propylene glycol, and glycol ethers. These solvents release volatile organic vapours into the air.

Can VOCs harm me?
Products that are sprayed from aerosol cans or pump sprays usually contain VOCs. Used in this way, VOCs become small particles that are easily inhaled and contribute to increased levels of ground-level ozone. These fumes can cause intoxication, drowsiness, breathing difficulties, asthma attacks, disorientation and headaches. Long-term exposure to some organic solvents can damage the nervous system and may be carcinogenic. Many solvents can be skin and eye irritants; others produce vapours that can be flammable. If you see the term 'volatile' on a label, ventilating the area you are working in is crucial to reduce harmful inhalation. Volatile compounds also release vapours slowly while being stored, thus creating toxins in the air of your home on a constant basis.

How do I know there are VOCs in my home?
You can usually smell VOCs easily – artificial fragrances contain VOCs – and the strong scents that come from many products such as glass cleaners are VOCs escaping from the product into the air of your home. Ironically, it is often the smell of VOCs that we associate with a 'fresh smell' in our homes. We feel certain that our homes are clean if they have this familiar scent. VOCs are also present in plug-ins and spray air fresheners.

Now that you are fully versed in the nasty chemicals that lurk in shop-bought cleaning products, let's learn about some safer alternatives.

Green ingredients

We've seen how some chemicals can be harmful to your health and the environment. This chapter outlines some safer household ingredients, many of which you may already have. For each ingredient listed you'll find out where to purchase it, a brief outline of how to use it and whether there are any precautions you should be aware of. As you proceed through the book you'll find specific recipes for using these ingredients in your own homemade household cleaners.

Baking soda

What is baking soda and where can I buy it?
Baking soda ($NaHCO_3$) is a fine white powder that can be found in the baking aisle of most grocery stores. This one ingredient can be used for cleaning almost anything, including worktops, sinks, the bathtub and hob.

How do I use baking soda?
Baking soda is a versatile, all-purpose, non-toxic, mildly alkaline cleaner. It cleans, deodorizes, scours, polishes and removes stains. Baking soda can help dirt and grease to dissolve in water, so is very effective in the kitchen to clean worktops, sinks and tiles. It makes an effective oven and hob cleaner, cleans burnt saucepans and will bring a shine to bath tubs and sinks. It can even help freshen your drains and toilet bowl.

In the laundry, baking soda eliminates perspiration odours, softens fabrics and removes some stains. Its deodorizing properties make it useful to sprinkle on carpets and upholstery before vacuuming, and on pet bedding.

To prevent or neutralize odours, place a small bowl of baking soda in the refrigerator or sprinkle in the bottom of a clean, dry bin. Baking soda absorbs odours from the air in your home, making a useful air freshener, so why not place a few small bowls of it around your home, to keep it smelling clean? Baking soda will clean and polish aluminium, chrome, jewellery, plastic, porcelain, stainless steel and even silver without scratching.

Is baking soda safe?
Baking soda is non toxic to humans and safe to use on most surfaces and fabrics.

Borax

What is borax and where can I buy it?
Borax is a natural mineral compound. It is a white powder and can be found along with the laundry supplies in supermarkets or ordered from pharmacies.

How do I use borax?
Borax is an effective deodorizer with antibacterial properties. It helps inhibit the growth of mould and mildew, removes stains such as grease and ketchup, and can be used as a laundry booster. It is a useful disinfectant and all-purpose cleaner.

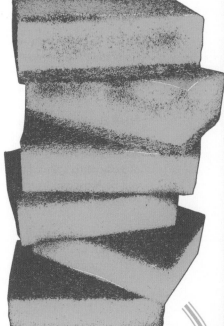

Mixed in equal quantities with baking soda and salt, borax makes an effective scouring powder. When dissolved in water, borax will loosen grease and stains from many fabrics. Borax also removes tea and coffee stains. Use it to deodorize the toilet bowl and in the dishwasher to clean dishes. Sprinkle it neat into a clean, dry bin to deodorize it and use on mattresses or carpets if there is a urine stain.

Is borax safe?
Borax is harmful if ingested, so don't use borax around food and keep it out the reach of children and pets. Borax can be a skin and respiratory irritant, so wear rubber gloves when using it and make sure you don't inhale the dust. In high doses, borax is toxic, so use with caution and reserve for heavy-duty cleaning only. Used occasionally with adequate care, it makes a useful ingredient for homemade cleaning products.

Club soda

What is club soda and where can I buy it?
Club soda is plain water into which carbon dioxide gas has been added.
It is an inexpensive drink found in the soft drinks section of any supermarket.
You can also make it yourself with a refillable seltzer bottle.

How do I use club soda?
Club soda's best use in household cleaning is to remove stains. It is well worth
keeping a bottle of this in your cupboard for emergencies. Pour club soda
directly onto stains and watch them fizz away. Blood, red wine and grease
will be loosened from fabrics, upholstery and carpets. It can save a carpet
or tablecloth from the fate of a red wine stain!

Many people use club soda for cleaning porcelain sinks, worktops and fixtures.
You can use it to clean the inside of your refrigerator, too. The beauty of this
product is that it doesn't need rinsing, so you'll save a lot of water, and
the environment will benefit from your frugal ways. Transfer some
club soda into a spray bottle to use on worktops for ease of
use. Another little-known use of club soda is on rusty nuts
and bolts. Pour club soda over them and the carbonation
will help bubble the rust away.

Is club soda safe?
Club soda is safe enough to drink! In addition, your
house plants will love a drink of flat soda water.
They'll drink it up happily and benefit from the
minerals.

Essential oils

What are essential oils and where can I buy them?
Essential oils are obtained from the fruits, leaves, flowers and resins of plants. They have powerful scents and unique qualities. They can be bought from health food shops, pharmacies and some supermarkets.

How do I use essential oils?
Some essential oils are antibacterial, anti-fungal or antiseptic. Tea tree, lemon and eucalyptus are powerful antibacterials that can be added to your homemade products. Other fragrances such as lavender, rose and orange are relaxing and invoke a sense of calm.

Smell is a personal thing, so spend time discovering scents you like then your housework can become a more pleasurable task. Although they can be expensive, you only need a drop or two. A small bottle can last months, even years!

Are essential oils safe?
You do have to exercise caution with essential oils. A few drops of most oils are safe for all the family, but in larger quantities they can be toxic. Essential oils should never be ingested and, apart from lavender and tea tree, you should not apply them neat to the skin.

Hydrogen peroxide

What is hydrogen peroxide and where can I buy it?
Hydrogen peroxide (H_2O_2) is a weak acid, readily available at 3 and 6 per cent dilution in pharmacies. All recipes in this book call for 3 per cent H_2O_2.

How do I use hydrogen peroxide?
H_2O_2 has strong oxidizing properties, so is useful for bleaching stains from white garments or removing stains and spills from light-coloured carpets and furnishings. It is particularly effective on blood. It can be used on most fabrics safely but is especially useful on white fabrics that won't suffer from its bleaching effect. H_2O_2 can be used to flood out urine stains on mattresses and to soak nappies before washing.

It is a useful disinfectant with some antibacterial properties. This makes it good for sanitizing surfaces such as worktops, cutting boards, light switches and toilet flush handles. Spray 3 per cent H_2O_2 onto the floor for use with a microfibre eMop or regular mop, and squirt it around the rim of your toilet and leave to work overnight. It can be used on tiles and around the shower to remove mould and mildew.

Is hydrogen peroxide safe?
Even when used at 3 per cent, hydrogen peroxide can be mildly irritating to the skin and mucous membranes on sensitive individuals, so always wear rubber gloves. Avoid splashing it on surfaces you don't want bleached, such as dark clothes. In high concentrations, hydrogen peroxide is an aggressive oxidizer and will corrode many materials and damage human skin. Due to its bleaching properties, always check surfaces and fabrics for colourfastness before applying H_2O_2.

Lemons

What are lemons and where can I buy them?
Lemons are citrus fruits that are readily available from grocers, farm shops and supermarkets. If you cannot get fresh lemons, buy bottled lemon juice instead.

How do I use lemons?
Lemons are acidic and the juice will cut through greasy surfaces whilst leaving a clean, fresh fragrance – add a squirt to your dishwashing water. To clean all types of floor, even wooden ones, add the juice of half a lemon to a bucket of hot water.

The antibacterial effects of lemons make them perfect for sanitizing chopping boards and other surfaces such as worktops or hobs. Add the juice of one lemon to a medium bowl of hot water for cleaning large areas, and rub the cut surface of half a lemon over smaller areas.

The mild bleaching power of lemons helps remove stains on clothes. Use the juice of half a lemon in a bucket of cold or warm water as a pre-soak for washing whites. This is a great solution for grubby nappies. In addition, the acid in lemons helps to break down limescale. Rub cut lemons around taps to descale them. Mixed with a little oil, lemon will bring a shine to wooden furniture adding a delightful fragrance.

Is it safe to use lemons?
When using lemon juice, make sure you don't get any in your eyes, otherwise it will sting. Although it won't harm you, lemon juice on a cut on the skin will smart for a while, so wear rubber gloves.

Liquid castile soap

What is liquid castile soap and where can I buy it?

True castile soap is made from olive oil and is one of the purest soaps available. Liquid castile soap can be purchased online or can be found in health food stores and is available pre-scented with essential oils.

How do I use liquid castile soap?

Buying peppermint- or eucalyptus-scented castile soap is particularly good for cleaning your home with its fresh, clean smell. Lavender-scented soap has relaxing, calming and soothing properties.

Castile soap makes a great base for all-purpose cream cleaners. It is an effective cleaner, particularly in soft water areas. In hard water areas, soap tends to leave a soap scum, which can leave a dull gray haze on some surfaces. A 50/50 mix of white vinegar and cold water removes soap scum.

You can use liquid castile soap in place of dishwashing liquids; one scented with lemon essential oil would be particularly pleasant to use for this job. Liquid castile soap can also be used to clean sinks, the hob, tiles and worktops.

A squirt of liquid castile soap in a bucket of warm water can be used to clean the floor and, in addition, solid bars of castile soap can be grated and made into 'laundry gloop' (see page 85) for your own laundry liquid and powders.

Is liquid castile soap safe?

As with all soaps, you should not get liquid castile in the eyes and it should be thoroughly rinsed after prolonged contact with the skin.

Microfibre cloths

What are microfibre cloths and where can I buy them?

Microfibre cloths are made from polyester, which you might not think appropriate for an environmentally-friendly book, however, microfibre cloths and mops keep for years. You can buy them in most supermarkets and hardware stores.

How do I use microfibre cloths?

The most amazing thing about these cloths is that you don't need any detergent or cleaning product to use them. All but the dirtiest jobs can be done with a microfibre cloth without the use of any toxic or homemade products. This will save you money and help protect your health and the environment.

These cloths can hold up to seven times their own weight in water, making them invaluable for spills. They have an exceptional ability to hold oils, so are perfect for cleaning the kitchen.

Microfibre is lint-free, so brings a lovely shine to mirrors and glass. The cloths can be safely used on TV screens and computer monitors and can withstand vigorous cycles through the washing machine. It's best not to use fabric conditioner with microfibre as this will affect the absorbency of the cloths.

Olive oil

What is olive oil and where can I buy it?

Olive oil, readily available in supermarkets, is obtained from the fruits of the olive tree. You don't need to spend a fortune on cold pressed oil for cleaning the home; buy the cheapest bottle and save the best for cooking!

How do I use olive oil?

Olive oil is very nourishing and makes a good ingredient for wooden furniture and floor polishes. The secret is not to use too much and to add an acid like white vinegar so that the surface you are polishing doesn't become so oily it attracts dust. This can leave you with a worse mess than you started with! Using too much oil can result in a surface that becomes highly slippery, which is not safe for floors, so test on a small area first.

A few drops of olive oil on a soft cloth will clean and shine stainless steel. You can also use it on leather shoes of any colour as a polish. If you prefer, use jojoba oil in any of the recipes that call for olive oil. Jojoba oil is a natural liquid wax that never goes rancid (unlike vegetable oils). It comes from the mature seed of a shrub and can be purchased in health food shops or online. It is expensive, but a little goes a long way.

Is olive oil safe?

Olive oil is safe enough to eat and is fine to get on your skin!

Salt

What is salt and where can I buy it?

Salt is a cheap, easy-to-find product and can bought in bulk from most supermarkets. You don't need to buy special salt for cleaning; table salt is ideal.

How do I use salt?

With its amazing ability to attract moisture, salt can be useful for 'pulling' stains from fabrics and carpets. If you spill red wine on a tablecloth, immediately cover the stain with salt to absorb the wine. A pre-soak in salt water will help remove perspiration stains on laundry. Combined with equal quantities of baking soda and borax, salt becomes a wonderful all-purpose scouring powder. If you dip a cut lemon in salt, you get an instant, safe and effective scouring pad to use on copper saucepans and chopping boards.

Rubbing tea and coffee stains with salt will remove them effortlessly. Oven and hob spills can be liberally sprinkled with salt and then brushed away. You can use salt in pans too. Sprinkle with salt before cleaning if you've cooked a greasy meal such as a roast dinner or stir-fry.

Salt can be used in combination with baking soda and white vinegar to unblock drains – it scours as it goes down the drainpipe.

Is salt safe?

Salt can be eaten in small quantities and it is fine to get it on the skin. It can hurt for a while in small cuts and can scour the skin a little which is uncomfortable, but essentially harmless.

Soda crystals

What are soda crystals and where can I buy them?

Soda crystals (sodium carbonate) can be found in the laundry section of supermarkets or ordered from pharmacies.

How do I use soda crystals?

Unlike many conventional laundry products, soda crystals contain no phosphates, enzymes or bleach. They are excellent at removing grease, red wine, grass stains, blood, tea and coffee stains, and ground-in dirt. Use on hobs, cooker hoods and grill pans. They add sparkle to tiles and can be used to clean the toilet. You can use them on shower curtains to remove mildew. A strong solution of soda crystals will also clear a blocked drain. Though they will effortlessly clean grease and stains, they can be too powerful for polished and lacquered surfaces, so be cautious.

Soda crystals are good for softening water, and pre-soaking woollens in a cold solution of soda crystals will leave them soft and fluffy. As a guide, use 1 tablespoon of soda crystals per 600 ml (1 pt) of cold water.

Are soda crystals safe?

Soda crystals should never be used on aluminium surfaces as they pit the metal and eventually turn your aluminium into 'lace'. However, this strong reaction can be used effectively to pull the tarnish from silverware. With their ability to cut through grease and dirt you must wear rubber gloves to protect your hands when handling soda crystals and keep them away from your face and eyes.

White vinegar

What is white vinegar and where can I buy it?
White vinegar is a traditional cleaning product that has been used for centuries and can be found in the condiments section of supermarkets.

How do I use white vinegar?
White vinegar is mildly acidic and useful for cleaning windows, mirrors and glass. A spray bottle filled with a 50/50 mix of cold water and white vinegar is an effective multi-purpose cleaner that will dissolve greasy marks, add sparkle to surfaces and remove residue of other cleaning products.

White vinegar has de-scaling properties, so is effective in removing limescale from tiles, kettles and taps (but not plated ones). Combining baking soda (which is a mild alkaline) with white vinegar (a mild acid) creates a 'fizzing' action which is useful for clearing blocked drains.

The smell of vinegar quickly dissipates and it is a deodorizer. Your home will not smell of vinegar for too long! As you work with it, vinegar freshens the air and neutralizes odours, leaving your home fresh and clean. White vinegar can also be used to remove underarm odours from clothing.

Is white vinegar safe?
As with all acids, keep white vinegar out of the eyes and any cuts on your skin. White vinegar is not dangerous, but can sting. Malt vinegar is not suitable for cleaning as the smell lingers and the colour can stain surfaces over time.

Getting started

In this chapter you'll learn to make your own household cleaning recipes. It features a number of basic recipes that you can use throughout the house, such as all-purpose cleaners, air fresheners, antibacterial cleaners and glass cleaners. You probably regularly use these products already, so why not make your own, eco-friendly versions of them? You'll never want to go back to shop-bought products again!

Air fresheners

What are air fresheners?

Air fresheners come in aerosol cans, pump sprays, gels, plug-ins or as scented candles. They mask unpleasant odours and give your home a 'just cleaned' scent.

How do air fresheners work?

Air fresheners work by coating the inner walls of the nose with a chemical to block the olfactory nerve, or they overpower the scent you want to mask with another potent fragrance.

What's in air fresheners?

Typical air fresheners contain ethanol, water and gases to propel the liquid out of the aerosol, as well as artificial fragrances.

Dangers to you

One brand states: 'Do not breathe spray'. It goes on to say: 'Deliberately inhaling the product may kill'. Surely they're designed to smell good, so you'd want to take in a lungful of air once you've sprayed it around your home? According to an article published in *New Scientist* magazine in 1999*, in homes where aerosol sprays and air fresheners are frequently used, mothers experienced 25 per cent more headaches and were 19 per cent more likely to suffer from depression. Infants under the age of six months had 30 per cent more ear infections and a 22 per cent higher incidence of diarrhea.

Dangers to the environment

One air freshener states on the label 'Harmful for you and the environment'. Additionally, the propellant gases used in aerosol cans are highly flammable and can be hazardous.

Keeping it eco-friendly

The first thing to understand about bad smelling air is that it is a symptom of something else, therefore you need to eliminate the root cause and not cover up the smell. So set your inner detective to work!

* *New Scientist* magazine, Issue 2202, 04 September 1999, page 17.

Around the house

Kitchen

Some smells are obvious; if you've been boiling cabbage in the kitchen it's clear where the smell is coming from. When cooking pungent foods, make sure you open kitchen windows or switch on the extractor fan if you have one, and close the kitchen door. Once you stop cooking the smells will dissipate and you'll be back to your sweet smelling home in no time.

Bathroom

If the toilet consistently smells bad, even when it's been cleaned, there may well be a blockage in the toilet that needs sorting. Alternatively, someone could have missed the toilet and the floor may need cleaning. Is there adequate ventilation in the room?

Laundry room

If there is a nasty smell in the laundry room, check for blockages in the hoses from your washing machine. Check the filters are clean and make sure there isn't a mouldy sock hiding somewhere.

Children and pets

If your dog smells, does it need more regular bathing? More exercise? A change of diet? Or a visit to the vet? And if your children's bedrooms stink like a farmyard, pull back the bed covers and check for wetting accidents. It's also worth hunting under the bed for forgotten secret stashes of food, and look on bookshelves and in cupboards for mouldy teacups or lunch boxes. Remember, there is always a reason for bad smells! If your home smells damp, or you have mould or mildew growing, there are ways to treat this instead of covering up the smell with air fresheners. See page 98 for suggestions.

Apply basic hygiene

Mopping spills when they happen, regularly emptying and cleaning bins, airing rooms and keeping the toilet spotless will all help to keep your home looking clean and smelling fresh. A freshly aired and clean room should not smell and there shouldn't be any need to artificially fragrance a room.

Creating naturally sweet smells

If you are taking baby steps towards a greener, cleaner home and like your home to smell of something fragrant, then there are plenty of safe and effective alternatives to the toxic chemicals found in many conventional air fresheners. Here are a few ideas.

Fresh air. This first solution is free! All you have to do is throw open your windows and let in some fresh air. Air pollution is two to five times higher inside homes than outside, even in inner city areas.

Baking soda is a natural neutralizer of nasty smells. Leave small dishes of it dotted around your home to absorb smells. Other neutralizers include undiluted white vinegar (use it in a similar way to baking soda by placing small cups of it around the house) and onions, cut in half with the peel left on. Unlikely as this may sound, your home won't end up smelling of vinegar or onions – it will smell fresh!

Scented candles can provide beautifully fragrant aromas, but don't replace one problem with another. Choose natural beeswax or soya candles that are fragranced with natural oils, not artificial fragrance or parfum. Alternatively, light an unscented candle and add a few drops of your favourite essential oil into the melted wax around the wick.

Ten ways with essential oils

Choose essential oils to match your mood or need: lemon, tea tree and eucalyptus are cleansing and purifying. Lavender, bergamot and grapefruit are uplifting. Geranium, lavender and frankincense will help you unwind and relax. Need to concentrate for exams? Try basil, rosemary and grapefruit. Ylang ylang, rose and sandalwood will set the mood for a romantic evening.

1. The simplest way to use essential oils in your home is with a plant mister. Use 8 drops of essential oil per 200 ml (7 fl oz) of water in the following recipe: Put 2 teaspoons of vodka in a plant mister. Add your chosen essential oil(s). Top up with cold water and shake well. The vodka is not essential, but it will keep the oils smelling fresh for longer and will help them to disperse.

2. You can use white distilled vinegar in a similar way. Half fill a plant mister with white vinegar. Add your chosen essential oil(s). Top up with cold water and shake well. The smell of vinegar will soon dissipate and you will be left with the beautiful and natural aromas of essential oils. Don't spray near silk or velvet, as discolouration may occur.

3. If you prefer a gel air freshener, it's easy and fun to make your own. Boil 115 ml (4 fl oz) of water and dissolve a packet of gelatin into it. Remove from the heat and add another 115 ml (4 fl oz) of cold water to the mix and stir until blended. Add a few drops of vodka (to act as a preservative) and ½ teaspoon of your chosen essential oil, and stir until well blended. Add a few drops of natural food colouring at this stage if you so wish. Pour the mix into pretty glass containers and allow to set. Dot the gel fresheners around the house but don't place them in direct sunlight or near a source of heat as they may start to liquefy! If this happens, simply place them in the fridge until set again.

4. Buy an oil burner and create your own home moods. Fill the receptacle with warm water, add around 6 drops of your chosen essential oil and light a tea light underneath. As the water warms, it evaporates to scent your home. Don't let it burn dry; top it up with warm water as necessary.

5. Electric oil burners are small, flat 'dishes' that come with an electric lead. Add a couple of drops of your chosen essential oil to the plate (don't add water to electric oil burners, just the essential oil). Plug in the device and switch on. It generates a small amount of heat to diffuse the fragrance into your home.

6. If you don't have any essential oils, boil 1 litre (1¾ pt) of water with half a dozen cloves, a couple of cinnamon sticks, the peel of two apples and an unpeeled, sliced orange. Allow to simmer and soon your home will smell delicious!

7. During the winter, put 2 drops of your chosen essential oil on a cotton wool ball and tuck it behind the radiator to gently infuse your home with a delicate scent.

8. You can buy especially designed light bulb rings for evaporating essential oils. Add a couple of drops of your chosen oil to the ring, and as the light creates heat, it will diffuse the scent into the room.

9. Why not buy or make herbal sachets to scent your home? Lavender bags are a popular choice which can be easily made with scraps of fabric and filled with dried lavender flower heads. Other herbs such as rosemary, thyme and lemon balm can be dried and used too.

10. If you are fortunate enough to have a woodburner, you can place a small, heatproof bowl filled with warm water on top of the stove. Add 2 to 4 drops of your favourite essential oils to the water. As the water evaporates, your chosen scent will fill the air.

All-purpose cleaners

What are all-purpose cleaners?
All-purpose cleaners are creams, liquids, sprays, powders and more recently, disposable wipes. They cover a wide variety of cleaning tasks, such as cleaning worktops, tables, kitchen hobs, bathtubs, sinks and even the floor.

How do all-purpose cleaners work?
All-purpose cleaners work to loosen grease and grime, so that you can rinse it all away.

What's in all-purpose cleaners?
All-purpose cleaners are made up of detergents (the cleaning bit), solvents (these help to dissolve some of the other ingredients and act as antifreeze so the products may be stored in cold places), fragrances, colours and preservatives. In addition, the powder cleaners contain abrasives to scrub away dirt. These can be made from bits of plastic or minerals found in the earth's crust, such as silica. Some all-purpose cleaners contain bleach or ammonia and others contain phosphates.

Dangers to you
Spray cleaners are a watered down version of a cream cleaner in an easy-to-use spray bottle. Unfortunately, using spray cleaners increases your risk of inhaling chemicals, as a fine mist is produced whenever you use them. Cream and powder cleaners are usually more 'effective' because they contain ammonia or bleach to increase their cleaning power. Unfortunately, these products can be irritating to the eyes, throat and lungs. If you use powder cleaners there is a risk that the minute particles of powder can permanently lodge in your lungs.

Dangers to the environment
Some all-purpose cleaners contain phosphates which are an environmental hazard. Disposable wipes place a large burden on the landfill. They are thrown away after one use and most come in non-recyclable plastic containers.

Keeping it eco-friendly

By adopting regular, basic hygiene principles in your home, you can drastically reduce, and even eliminate, the need for toxic cleaning products. Wipe the hob and oven after each use to prevent grease build-up; clean over the worktops after any meal preparation to prevent grime; keep a spray of neat white vinegar (or a 50/50 mix of white vinegar and cold water if you prefer) in the bathroom to use after a bath or shower to prevent soap scum build-up; and wipe down doors and cupboards to prevent oily fingerprints.

Eco-friendly alternatives

Here are some effective and safe alternatives to all-purpose cleaners.

Water. It's our old-fashioned friend, hot water! Hot water won't cut through grease, but it will clean everyday spills effectively so that grease and dirt don't build up so easily.

A steam cleaner is a worthwhile investment. It has the added benefit of killing germs. You can use steam to clean your kitchen, bathroom and to sanitize worktops. A good quality steam cleaner will remove baked-on grime and grease from microwaves, hobs and ovens. In the bathroom, you can use it to clean and sanitize taps, shower curtains, toilet bowls and shower cubicles.

Baking soda is excellent if you are used to a powder cleaner. Use it on the hob, oven, bathtub, sink and shower, on tiles, worktops and cupboard doors. Baking soda can leave a white, powdery residue, so rinse with plenty of warm water or spray a 50/50 mix of white vinegar and cold water on it afterwards to remove any remaining film. Sprinkle a little baking soda directly onto a damp cloth and use to gently scrub grease, grime, food spills and tide marks away from around the bath, washbasin and sink. Baking soda works wonders on the inside of fridges as it deodorizes while it cleans.

If you're after a more heavy-duty scouring powder, then the following will work wonders. Put on a pair of rubber gloves and mix a cupful each of borax, baking soda and cooking salt to make an effective multipurpose cleaner. Store in a jar with a well-fitting lid. Apply to surfaces and scrub with a damp cloth or sponge.

General all-purpose liquid cleaner. Fill a spray bottle one-third full with white vinegar. Top it up until it is nearly full with cold water. Add 10 drops of your favourite essential oil (lemon, eucalyptus, orange or tea tree are great for the kitchen and bathroom). Works on tiles, mirrors, worktops and appliances; you can even spray and leave it on with no need to rinse! Add 100 ml (3½ fl oz) of this solution to a bucket of hot water for cleaning hard floors.

Borax can be used to clean a really tough, greasy area, like a cooker hood. Remember to wear rubber gloves when using borax, don't inhale it and don't use it near food. It's safer than many conventional products, but it is still toxic. To clean a cooker hood filter, add 125 g (4½ oz) of borax to a sink of hot water. Leave the filter to soak for 15 minutes and rinse well.

A heavy-duty spray can be used for cleaning very greasy areas, such as tiles, the back of the hob or the tops of kitchen cupboards. In a clean 500 ml (18 fl oz) spray bottle, mix 30 ml (1 fl oz) white vinegar with 1 teaspoon of borax. Add 450 ml (16 fl oz) of hot water and shake. Add ½ teaspoon of liquid castile soap and 5 drops of essential oils (3 of lemon and 2 of lavender is nice). Shake before use and spray onto surfaces, then rinse off.

Soda crystals are another effective way to clean greasy areas. Using the cooker hood example again, use 200 g (7 oz) of soda crystals in a sink of hot water with a tiny squeeze of hand dishwashing liquid – not too much or you will get too much foam! Put on rubber gloves and soak the cooker hood for 15 minutes and then rinse. Do NOT use on aluminium surfaces. If the cooker hood is aluminium, use borax instead.

Make up a solution of 1 tablespoon of soda crystals to 600 ml (1 pt) of cold water for a maintenance spray and use this weekly on the exterior of your cooker hood or other non-aluminium areas to stop grease building up.

All-purpose cream cleaner. In a container with a lid, mix 90 g (3 oz) of baking soda with 115 ml (4 fl oz) of liquid castile soap and 6 drops of your favourite essential oils (optional). Use it to clean sinks, bathtubs, tiles, appliances or anywhere that you want to keep clean and fresh. Shake before each use. It brings a wonderful shine to bathtubs and sinks.

Antibacterial cleaners

What are antibacterial cleaners?

Antibacterial cleaners kill germs in your home. You'll find antibacterial cleaners for the kitchen, bathroom, floors, worktops and in dishwashing detergents. Even some food storage boxes and cutting boards come impregnated with them.

How do antibacterial cleaners work?

Antibacterial cleaners contain chemicals which kill bacteria and microbes (but not viruses). They come as liquids, sprays and wipes. Antibacterial worktop cleaners are usually products you spray onto a surface and leave on, without rinsing. Liquids are generally added to a bucket of water and used to clean large areas, such as floors. Wipes are once-use disposable items that are used on smaller areas like tables, highchairs and light switches.

What's in antibacterial cleaners?

Antibacterial cleaners contain detergents, solvents, fragrances, preservatives and colours, plus an antibacterial agent. One of the most common antibacterial ingredients is triclosan.

Dangers to you

Growing evidence reports that antibacterial products may do us and the environment more harm than good. The biggest concern is that we are seeing an increase in antibiotic-resistant organisms or 'superbugs'.

For most people, strong antibacterial cleaners are unnecessary. Coming into contact with germs as a healthy individual doesn't cause problems. However, if there are elderly people, those with compromised immune systems or newborns in your home, the use of safe, natural antibacterial cleaners can be an advantage for keeping germs at bay. The immune system is a bit like a muscle: it needs a regular workout to keep it strong, healthy and functioning well. In addition, most common infections such as coughs, colds and flu are caused by viruses, not bacteria. So antibacterial cleaners won't make any difference anyway!

Triclosan is a popular antibacterial ingredient added to conventional products such as household dishwashing soaps. Reports have suggested that triclosan can combine with chlorine in tap water to form chloroform gas, which the United States Environmental Protection Agency classifies as a probable human carcinogen. The common sense approach, therefore, is to use antibacterial cleaning products only when there is a real need for them. Imagine antibacterial cleaners to be like medicine. You only take medicine when you're ill. You wouldn't dream of taking cough medicine or headache tablets on a daily basis.

Dangers to the environment
In the environment, triclosan may react with sunlight, forming other compounds which may include chlorophenols and dioxin. Although small amounts of dioxins are produced, there is a great deal of concern over this effect because dioxins are extremely toxic and very potent endocrine disruptors.

Keeping it eco-friendly
If this has put you off using conventional antibacterial products, but you're still concerned about germs lurking in your home, there are plenty of safe, natural alternatives. One of the most important steps to prevent bacterial build-up in your home is to keep surfaces *dry*. Bacteria need a moist environment to thrive, so drying surfaces after you have cleaned them means bacteria cannot multiply.

Combine this with other preventative measures, such as mopping up spills, keeping the toilet hygienic, regularly washing dishcloths and keeping chopping boards clean (especially those used for raw meat). In addition, hang tea towels and cloths to dry after use and wash your hands thoroughly after visiting the toilet and before meal preparation.

If someone in your home has a stomach bug for example, then use a homemade antibacterial spray to wipe down taps, the toilet flush handle, doorknobs and light switches. You can also spray antibacterial essential oils into the air to purify and cleanse the air.

Homemade recipes

Essential oils

Combine the following essential oils in a spray bottle with 200 ml (7 fl oz) of cold water: 2 drop of tea tree oil, 1 drop of lavender oil, 3 drops of lemon oil and 2 drops of orange oil. Spray directly onto surfaces, wipe over with a damp cloth and dry with a clean cloth.

Hydrogen peroxide

Spray areas with a 3 per cent solution of hydrogen peroxide. At high concentrations, hydrogen peroxide can be dangerous. Used at 3 per cent, however, it is safe and effective.

Borax

Borax has long been known for its disinfectant and deodorizing properties. It does have toxic properties, but if there are germs in your home, or there is someone in your home with a compromised immune system, borax can be helpful for disinfecting surfaces.

To make a simple cleaning solution for any type of floors and surfaces, wear rubber gloves and combine 125 g (4½ oz) of borax with 4.5 litres (1 gallon) of hot water. For a convenient spray, mix 4 tablespoons of white vinegar with 2 tablespoons of borax in a 500 ml (18 fl oz) spray bottle. Top up with very hot water and agitate the mix to dissolve the borax. This spray is ideal for door plates, toilet handles and around taps.

And now for a story...

During the Bubonic plague, four thieves were captured and charged with robbing the dead. The court offered a lenient sentence if they would share how they had managed to avoid the plague whilst having close contact with the dead and dying. The thieves revealed that they were perfumers and spice traders. They told of a blend of aromatic herbs, including cloves, cinnamon and lemon which they had rubbed over their bodies to protect themselves. Some companies now offer a 'four thieves' blend of essential oils, which is claimed to be more effective against bacteria than harsh chemicals and drugs. So another alternative to toxic antibacterial products is to use a blend of four thieves essential oils!

Window, glass and mirror cleaners

What are window, glass and mirror cleaners?

Window, glass or mirror cleaners are sprays you wipe onto surfaces and buff with a cloth. They also come as disposable wipes which you wipe over surfaces before discarding.

How do window, glass and mirror cleaners work?

These cleaners contain ingredients that cut through grease and grime. Grease builds up on glass surfaces to leave marks and a hazy appearance. Window, glass and mirror cleaners usually boast a smear-free finish.

What's in window, glass and mirror cleaners?

You will find ingredients such as solvents, sodium hydroxide (lye), naphtha, hydrochloric acid, artificial fragrances and colours.

Dangers to you

Most people use spray cleaners to clean their home. These create a fine mist of chemicals which are easily inhaled and can travel all over your home. They can rest on other surfaces such as the dining table where you eat. Solvents dissolve dirt and speed the evaporation rate of the product but are readily absorbed by the skin and have been associated with liver and kidney problems. Glass cleaners contain perfumes and preservatives. One commonly used preservative, methylisothiazolinone, has been shown to be allergenic and cytotoxic (toxic to cells) in studies. Another, methylchloroisothiazolinone, can cause chemical burns in high concentrations, so much so that it has been removed from most cosmetic products. Artificial fragrances can cause asthmatic reactions, allergic skin reactions, nausea, disrupt the balance of hormones in the human body and even cause cancer.

Dangers to the environment

Disposable wipes are a once-use product and many people dispose of them in the wrong way by flushing them down the toilet. Disposable wipes should be put into the landfill, but here they may remain for many years, leaching out small amounts of chemicals into the environment.

Keeping it eco-friendly

Use e-cloths or old rags rather than disposable paper towels for cleaning glass. To get a perfect finish on glass, mirrors and windows, apply the product with one cloth and buff with a dry, lint-free cloth. Drying is the secret to a streak-free finish – keep buffing until the glass is completely dry!

Although the sun streaming through the windows on a spring morning makes us all want to go and clean our windows, this is the worst time to do it! Cleaning windows with the sun shining on them leads to streaks because the sun dries the cleaning solution too fast. Wait for the clouds to come so that you don't waste your energy or product.

One idea to save on the amount of energy you expend is to start cleaning the outside of your window with side-to-side movements. When you have done this, clean the inside with up-and-down movements. If there are any streaks, you'll know which side of the window they are on from the direction they are going. Clean your windows regularly as a maintenance part of your housework rather than leaving them until they are really dirty. Keep a spray bottle of white vinegar and water in the bath and shower rooms to regularly use on mirrors and glass shower doors. If you keep your windows and mirrors regularly wiped over, you'll find you can clean with plain water and an e-cloth which is the simplest formula!

Homemade recipes

Keep it simple with vinegar

The best cleaner for glass, windows and mirrors is white vinegar. Some conventional products even boast 'added vinegar for extra streak-free shine!' Mix a 50/50 solution of white vinegar and cold water in a spray bottle. Spray directly onto the glass or onto a soft cloth to remove grease and grime. If you use an e-cloth, you will be rewarded with an effortless streak-free finish. Add 4 drops of essential oil to the mix if you wish. Lemon and peppermint are invigorating and will leave a lovely fresh smell in your home.

Join the club

Plain soda water works wonders on glass. Pour in a spray bottle and use on windows, mirrors and glass. It takes longer to dry than vinegar, so keep working at it with a dry, lint-free cloth to get a streak-free finish.

Cutting through grease

If your windows or mirrors are really dirty and greasy, then you'll need a little more cleaning power. Many conventional products contain artificial waxes which coat glass, so don't be surprised if your first attempt at going 'natural' with window and mirror cleaning is not a success and your windows look worse than ever! Try this more 'heavy duty' formula to remove the wax layers.

In a spray bottle combine 2 parts cold water with 1 part vinegar and around $1/2$ teaspoon of liquid castile soap or an eco-friendly dishwashing detergent. Spray the mix onto your windows and clean with a squeegee mop, from top to bottom of your windows. Buff with a dry, lint-free cloth. You may need to repeat this a few times to get the windows really clean.

Soda crystals wonder

If the above solution doesn't work to cut through grease, then the following will give you windows your neighbours will envy! Wear rubber gloves and mix 1 tablespoon of washing soda with 600 ml (1 pt) of cold water. Wring out a cloth in the solution, wipe over the windows from top to bottom and clean with a squeegee mop. Buff with a dry, lint-free cloth.

Well done! You've learned about some basic cleaning recipes. Now we're going to move into the kitchen and learn how to keep this important room clean.

The kitchen

There are lots of products to keep our kitchens clean. There are all-purpose cleaners, special cleaners for ovens and drains, kettle descalers, antibacterial products and dishwashing detergents, to name but a few. These products are designed to keep our kitchens safe and clean, but many of them contain alarming toxins which should not be used anywhere near food. Fortunately, there are some natural and safer alternatives that will help you tackle the toughest cleaning jobs.

Chopping boards

It is important to keep chopping boards clean, especially if you cut raw meat or fish on them. Ideally, you should have one board for raw meat and fish preparation and another for different foods, such as vegetables, cheese and bread. You might prefer to buy a small board for cutting meat and fish which can be totally submerged in a bowl of hot, soapy water.

V is for vegetables

If you do not have two boards, then mark one of the sides 'V' for vegetables and the other side 'M' for meat. Prepare raw vegetables first, on the side marked 'V', then flip the board over to prepare meat. Once you have finished preparing your meal, sanitize both sides of the board.

Materials used for chopping boards

Most chopping boards are made from wood or plastic. Some are made from acrylic, silicone or glass, but these are usually worktop protectors and are not meant for prolonged use as chopping boards. There are good and bad points about each material.

Wooden chopping boards: wooden boards are usually heavier than plastic ones, will last for decades if well looked after and have natural antibacterial qualities. Wooden boards can be FSC certified, which means they are from sustainably managed forests and biodegrade at the end of their lives. Wooden boards need seasoning before their first use and regular re-oiling to keep them sealed against bacteria. They cannot be submerged in water because they can crack as they dry out, so must not be put into a dishwasher.

How to season a wooden chopping board
You will need to oil a new wooden board around five times to get it seasoned and ready for use. This will prevent staining, mould growth and absorption of odours and bacteria. After that, maintain the integrity of your board with a once-monthly application of oil. Do not use vegetable oils as these go rancid. Instead, use coconut oil or a food-grade mineral oil specifically intended for use with unseasoned wood. Before applying the oil, pour a little in a cup and warm it by placing the cup over a bowl of hot water. Apply with a soft cloth and allow to soak in for four to six hours before wiping away any excess oil. Repeat this five times over the course of a few days. It is worth investing in the best chopping board you can afford. Pine chopping boards tend to be quite soft, so they dent and scratch easily. A solid hardwood board, such as one made from oak or ash may cost more, but will last you a lifetime.

Plastic chopping boards: plastic is lighter to manoeuvre around the kitchen, cheap to buy or replace and can be placed in a hot dishwasher to kill germs. However, it scratches easily and germs can multiply in these scratches if not properly sanitized. Plastic chopping boards do not degrade and are made from a non-renewable resource. Some plastic chopping boards are impregnated with antibacterial ingredients which can be harmful to people and the environment. See Antibacterial cleaners on page 48 for more information.

Glass boards: these are easy to keep hygienic and sanitized as they can be put in the dishwasher or submerged in boiling water. Glass boards are not really designed for a lot of chopping – they are meant more as worktop protectors, but occasional chopping will not damage them. Glass boards blunt knives easily and there is the risk of breaking them if you drop one.

Which is the safest material?
Many think plastic is more hygienic because it is easier to clean than wood. However, it has been scientifically proven that wooden boards are safer and more hygienic than plastic because they do not accumulate as many bacteria.

How to keep your chopping board hygienic

Whichever type of chopping board you choose, ensure it is kept clean and washed after each use to prevent the spreading of bacteria such as E. coli and salmonella. Scrub your chopping board well with hot, soapy water after each use and dry it thoroughly. Bacteria need a moist environment and can only survive a few hours without moisture. If you are preparing a meal that contains meat, then wash your chopping board in between preparing raw meat and any other parts of your meal, such as the vegetables. In preference, prepare vegetables first, and cut the meat last.

Basic hygiene rules

Remember to wash your hands before and after food preparation, and do not use the same knife for raw meat and other parts of the meal. Following these basic hygiene rules of keeping chopping boards, knives and hands clean will help you, your family and your home to remain healthy and safe.

Ways to sanitize your chopping board

Lemons and limes: wipe over your board with the cut side of a fresh lemon or lime. Alternatively, squeeze fresh lemon or lime juice over the board, leave for a few minutes, then wipe with a damp cloth.

Salt: use salt like a scouring powder to remove odours, clean and disinfect your boards. You can even pour some salt over the cut end of a lemon to make your very own scouring 'sponge'! Allow the salt to remain in contact with the board for around 10 minutes before rinsing off with very hot water. This solution will remove odours such as garlic, lemon or fish. Note that regular salting can dry out wooden boards, so you may need to re-oil them more regularly.

Hot water: fit a non-wooden chopping board into a bowl, pour boiling water over it, leave until you can handle it, then dry well.

White vinegar: vinegar has deodorizing properties. Keep a spray bottle of neat white vinegar for spraying on your boards after washing. Follow with a spray of hydrogen peroxide at 3 per cent strength.

Drain cleaners

What are drain cleaners?

Drain cleaners clear blockages in drains and kill germs. These products are used to dissolve grease and hair clogs. They are used in the drainpipes of sinks, bathtubs and showers. They are available as granules or liquids.

How do drain cleaners work?

The ingredients in most conventional drain cleaners work by literally eating away or dissolving grease and hair clogs that block the drain.

What's in drain cleaners?

The primary ingredient of most drain cleaners is sodium hydroxide ($NaOH$), or lye. Other conventional drain cleaners contain sulfuric acid. Products containing sulfuric acid are really intended for professional use but are legal to sell to the general public. Drain cleaners also contain artificial fragrances.

Dangers to you

Along with oven cleaners, drain cleaners are among the most hazardous household products available to the public. Sodium hydroxide is caustic, and in strong concentrations can seriously burn the skin, eyes and mucous membranes. If swallowed, lye will burn the esophagus and stomach or cause death. One drop of concentrated lye on the skin can cause severe burning and if it gets into the eyes, can cause blindness. According to safety data sheets, inhaling sodium hydroxide is harmful and can cause death. Sulfuric acid is very corrosive and hazardous if misused. Exposure to it causes serious burns and it is harmful when inhaled. If ingested, sulfuric acid can be fatal. Chronic exposure may result in lung damage and possibly cancer.

Dangers to the environment

Drain cleaners containing these toxic ingredients are detrimental to septic tanks because they kill off the friendly bacteria that enable products to decay in the tank. If you have a septic tank, it is vital that you use a more gentle method of removing drain clogs, otherwise your tank will not function properly.

Keeping it eco-friendly

There is a clear case for prevention rather than cure when it comes to keeping drains unclogged and fresh rather than having to resort to using such dangerous products! Throughout this chapter you will find recipes to clear clogs and preventative recipes to help keep drains clear.

Homemade recipes

To clear a smelly drain

Everyone gets smelly drains from time to time. Here are two natural and safe methods to freshen your drain.

- **White vinegar**: if your kitchen or bathroom drain smells, then freshen it with white vinegar. Pour a cup of vinegar down the plughole and leave it to work. Add 5 drops of your favourite essential oil to the vinegar before using if you wish. Do this once a week to keep things smelling sweet.
- **Baking soda**: neutralizes odours. Pour a cup of baking soda down the drain and leave to work for an hour before pouring a kettle-full of boiling water down there.

To prevent blocked drains

It is best to avoid toxic drain cleaners if at all possible. One way to do this is to prevent a drain becoming blocked in the first place. Here are some simple tips to keep things flowing freely.

- Scrape dishes well before putting them in the sink. This prevents food bits going down the plug hole into the drain where they may build up and create a blockage.
- Use a drain trap to prevent food bits or hairs going down the drain.
- Pour boiling water down the drain once a week to prevent build-up.
- Don't pour cooking oil or fat down the drain. Instead, wipe fat or grease from pans with kitchen towel and compost it. For larger quantities of oil, pour it into an old container and take it to the recycling centre. During cold winter months, smaller amounts can be put out for the birds, perhaps.
- Use a sink plunger once a week or a plumber's snake to keep drains unclogged.

• Washing soda is an
excellent choice for
helping drains stay clear
because is it alkaline and it
dissolves grease. Once a
week, pour ¼ cup of washing
soda down the drain, and flush
with hot water.

To clear blocked drains
If you're unfortunate enough to
end up with a blocked drain, the following ways will help to clear it without
resorting to toxic chemicals.

• **Take the plunge**: try a plunger first. You may have to work hard at it, but
sometimes this will clear a drain. Use a damp cloth to cover the overflow to
prevent anything squirting out of it onto you. Partially fill the sink with water
and plunge vigorously several times before pulling the plunger off the drain
opening. Repeat as necessary. Once the drain is clear, be sure to maintain a
weekly 'prevention' routine from the ideas above.

• **Kitchen chemistry**: if a plunger doesn't work, then it's time for a little
kitchen chemistry! Acids and alkalies tend to neutralize one another so you
can use them to clear your drain. Pour 1 cup of baking soda (alkaline) down
the drain, followed by ½ cup of salt and finally 1 cup of white vinegar (acid).
The acid and alkaline will fizz, bubble and expand, break down grease and
fats, and the salt will act as an abrasive, which may be enough to dislodge
the blockage and clear the drain. Leave this to work for 15 minutes before
pouring boiling water down the drain to rinse everything away.

As you are not using any toxic chemicals, it is safe to plunge this after letting
the ingredients work and the boiling water cool. You should never plunge a
drain after pouring a lye-based cleaner down there in case it comes back
up the drain and splashes onto you. If you cannot clear a clog after a few
attempts, then call a professional plumber or drain cleaning service. If you
keep putting too much force on a drain or pipe, you could cause serious
damage to your plumbing.

Automatic dishwashing detergents

What are automatic dishwashing detergents?

Automatic dishwashing detergents are powders, liquids or tablets which are specially designed to work with automatic dishwashers and are not to be used for hand washing dishes.

How do automatic dishwashing detergents work?

When you wash by hand, it is the physical scrubbing and scouring motion that removes grease and dried-on food. If you are expecting a machine to do the same job without any hands and fingers, it needs some powerful chemicals to make that happen!

What's in automatic dishwashing detergents?

Most automatic dishwashing detergents contain chlorine compounds that are activated when they come into contact with water, as well as enzymes, phosphates, perfumes and colourants.

Dangers to you

The high alkaline nature of automatic dishwashing detergents means that they can irritate skin, so keep them off your skin, out of your eyes and away from babies and children. If swallowed, dishwashing detergents will burn the mouth, throat and oesophagus. Conventional automatic dishwashing detergents can release chlorine fumes while the dishwasher is running, and the steam that comes from them when you open the door is full of chlorine. Chlorine can cause headaches, fatigue, sore throat, and irritated eyes and lungs when inhaled.

Dangers to the environment

Some conventional automatic dishwashing detergents contain a high concentration of phosphates. Phosphates are a major cause of algae blooms which are hazardous to fish, plant life and water quality.

Keeping it eco-friendly

To keep your dishwashing habits environmentally friendly, only run the machine with a full load and do not use a pre-rinse unless the contents are really greasy. Save on energy by opening the door of the dishwasher and air drying instead of running the drying cycle. Unfortunately, there are not many self-sufficient alternatives to these harsh cleaners. If you decide to stick with a conventional product, then at least cut down the amount of detergent recommended. You might find that you only need half the amount of cleaning product to produce sparkle and shine.

If you use a conventional detergent, then always run your dishwasher in a well ventilated room to prevent build-up of chlorine in your home. More and more companies are producing chlorine- and phosphate-free alternatives, so try an environmentally-friendly brand instead.

Maintaining the health of your dishwasher
Follow these tips to prolong the life of your machine.
• Scrape off food bits before putting plates into the dishwasher to prevent clogs, blockages and smells.
• Every month, run the dishwasher empty, with a cup of white vinegar in the detergent reservoir. This will keep the machine clean, deodorized and fresh, and will remove any limescale.
• Leave the door ajar slightly when the machine is empty to ensure good air flow and prevent musty smells.
• Clean the inside of the dishwasher with a spray of neat white vinegar every couple of weeks.
• Always keep the salt reservoir topped up as this regenerates the machine's built-in water softener.

Homemade recipes
DIY dishwashing detergent
Here are two homemade dishwashing detergent recipes:
1. Wear rubber gloves and combine equal quantities of baking soda and borax in a jar with a tight-fitting lid. Add 2 to 4 tablespoons of this mix to the detergent compartment and fill the rinse aid receptacle with white vinegar.
2. Add 2 tablespoons of washing soda to the detergent compartment and fill the rinse aid receptacle with white vinegar. One of my friends swears by this method and she lives in a hard water area!

Cleaning the refrigerator

What are refrigerator cleaners?
All-purpose, antibacterial cleaners or mild bleach solutions are commonly used to clean refrigerator shelves, seals and doors. There are a few specific products for refrigerators, such as deodorizers.

How do refrigerator cleaners work?
Most conventional fridge products are designed to eliminate or neutralize odours rather than clean the fridge.

What's in refrigerator cleaners?
All-purpose cleaners and antibacterial cleaners contain bleaches, solvents, formaldehyde and disinfectants, along with ingredients like triclosan.

Dangers to you
Triclosan and formaldehyde have been classed as probable human carcinogens. Bleach can be irritating to the skin, eyes and throat, and can cause severe stomach irritation if ingested. Products like these are best kept away from food to prevent ingestion. One well-known fridge deodorizer states 'Do not eat', on the packaging. Another contains perfume, which is made up of hundreds of synthetic chemicals. Do you want these products near your food?

Keeping it eco-friendly
Refrigerators can be energy-hungry appliances. Reduce the amount of energy your refrigerator uses by regularly cleaning the coils at the back of the appliance and dusting underneath. If the coils get clogged with dust, they cannot work as effectively. Cool air needs to circulate around the food stored in your refrigerator to keep it working efficiently so avoid cramming too much in.

Once a week, check through the food in your refrigerator and see if it is still safe to eat. Any fruit or vegetables that have gone off can be composted. If the food is fresh, your refrigerator won't smell bad, so keep an eye on anything lurking in the back of the shelves or in the vegetable tray that needs using up or discarding.

Keep raw meat on the bottom shelf to prevent it dripping over other foods. Mop up any food spills straight away to prevent staining and bacteria build-up. Wipe over the outside of the refrigerator daily with a cloth dipped in white vinegar to keep greasy fingerprints and food spills at bay.

Heavy-duty recipes

Borax: if you've inherited a second-hand refrigerator that has been neglected, you can clean and deodorize it with borax. This shouldn't be necessary for a well cared-for appliance, but if yours is old or has been left with rotting food in it, you'll need a tougher recipe.

Wear rubber gloves and mix 1 tablespoon of borax with 1 litre (1¾ pt) of warm water. Use this solution on a soft cloth or sponge to wipe both the inside and outside of the refrigerator and all the shelving and containers. Rinse well with warm water.

Washing soda: if it's really bad with greasy residue, then washing soda crystals will save the day. Ensure this solution does not come into contact with aluminium. If your refrigerator has aluminium parts, such as the cooling plate, use borax instead.

Wear rubber gloves and mix 1 tablespoon of washing soda with 600 ml (1 pt) of warm water. Use this solution on a soft cloth or sponge to wipe the inside and outside of the refrigerator and all the shelving and containers. Rinse well with warm water.

Ten steps to cleaning your refrigerator the safe way

1. Take the food out of your fridge and check the use-by dates. Discard anything that is not fit for consumption and menu plan around the items that are coming to the end of their useful life. Wilting vegetables can be made into soup or stock; tired fruit can be whizzed into a smoothie or made into pies.

2. Take out any removable shelves, racks, containers or baskets and remove any food debris. Place the refrigerator parts in a bowl of hot, soapy water and clean them. An eco-friendly dishwashing soap is ideal for this.

3. While the shelves and racks are drying, tackle the inside of the refrigerator. Clean the inside with baking soda. Dab a damp sponge or cloth into some baking soda and use it as a gentle cleaner. Do not use harsh abrasives or metal scouring pads or you will scratch the inside of your refrigerator.

4. Clear any debris from the drip tray drain hole. Once this is clean, ensure that no food touches the cooling plate at the back of your refrigerator. When this happens food is instantly frozen onto the plate. When the compressor turns off, the plate warms up, melts the ice and water drips off which washes any particles into the drip tray drain hole causing a blockage.

5. Rinse the inside of the refrigerator with hot water and spray with a 50/50 mix of cold water and white vinegar to remove any last traces of baking soda.

6. Wipe down the outside of the doors and the seals with baking soda on a damp sponge. Rinse with a 50/50 spray of cold water and white vinegar. Dry with a clean cloth.

7. Dry off all the shelving and racks, and put them back in the refrigerator.

8. Give any bottles and jars a wipe over with a damp cloth in case they are sticky, before putting all the food back into your refrigerator.

9. Place a small, open container full of baking soda in the refrigerator. This will help neutralize smells.

10. You might need to turn the refrigerator up for an hour or so, to get it quickly back up to temperature.

Products for bins

What are bin cleaners?
With our determination to eradicate germs and smells, there are cleaning and deodorizing products for our bins. You can buy antibacterial freshening gels, powders and even scented plastic liners.

What are these products for?
Antibacterial freshening products contain powerful biocides to kill germs and artificial fragrances to cover up bad smells. Scented plastic liners cover up smells with something more attractive.

What's in bin cleaners?
A typical bin deodorizing product contains biocides such as benzalkonium chloridem and artificial fragrances.

Dangers to you
Benzalkonium chloridem is an allergen, and several studies have cast doubt on its reputation for safety. Recommendations for some of these products include washing hands after use, avoiding contact with the eyes and getting medical attention if swallowed. Benzalkonium chloride has been shown to lead to swelling of membranes and rhinitis. Scented bin liners contain artificial fragrances, some of which have been linked to a wide range of conditions from skin, nose and eye irritation to asthma, headaches and dizziness.

Keeping it eco-friendly
You can now buy biodegradable bin liners from supermarkets and household stores. These bags are usually made from some kind of starch, such as rice or potato. They save a build-up of plastic in the landfill because they decompose more quickly and are not made from oil, like most conventional plastic liners.

Think about where you can reuse other bags for your bins. Large rice sacks, old carrier bags or potato sacks are all ideal. Look at where you can reduce, reuse and recycle to put as little as possible into the landfill. Could you start recycling more or switch brands to products that come in less non-recyclable packaging? The less rubbish you create, the less maintenance your bins need!

If you rinse your bin out every time it is emptied with a mild solution of dishwashing liquid, a 50/50 cold water and vinegar mix or a mild soda crystals solution, this should prevent bad smells and bacteria building up. Try to keep liquid or oily waste out of your bins by reducing food waste so that your bin won't get so messy and dirty.

Homemade recipes

Indoor bins

To clean your indoor bins, make up the following solution: Put on rubber gloves and dissolve 125 g (4½ oz) of washing soda crystals into 600 ml (1 pt) of cold water. Use this solution with an old rag to clean your bin then leave to air dry.

Outdoor bins

For outdoor bins that are particularly greasy or dirty, make the following stronger solution: Put on rubber gloves and dissolve 250 g (9 oz) of washing soda crystals in 600 ml (1 pt) of cold water. Use this solution with an old rag or mop to clean your bin and leave to air dry.

Deodorize your bins with baking soda. Sprinkle a small amount of baking soda into a dry bin to neutralize odours every time you clean it out.

Kettles, coffee makers, teapots and cups

Kettles and coffee makers
Kettles and coffee makers can end up with a build-up of limescale, particularly in hard water areas. Limescale can impair the flavour of your drinks and cause blockages in the small pipes of coffee and espresso makers.

Using white vinegar is a safe and easy way to break down limescale deposits. In a kettle, cover the element with white vinegar, bring to the boil, then leave to cool. It's a good idea to do this job before you go to bed so that you don't have to go too long without your caffeine fix! In the morning rinse the kettle thoroughly before using. In a coffee maker, run a cycle through the machine with 60 ml (2 fl oz) of white vinegar topped up with water in the water reservoir. Follow with a cycle of plain water to get rid of the vinegar taint. Limescale can build up to such an extent that it prevents kettles and coffee makers working, so use white vinegar whenever you see limescale building up.

Teapots and cups
These regularly have brown stains from tannin in tea or are stained from coffee. Many people use a mild bleach solution to clean teapots and cups. There is no need to resort to harsh bleach as salt and white vinegar will work effectively! Mix some salt and white vinegar into a paste. Wipe this paste around the stain with a soft cloth and rinse thoroughly.

For really badly stained cups or teapots, use a strong solution of soda crystals. Put on rubber gloves and mix 250 g (9 oz) of soda crystals with 600 ml (1 pt) of cold water. Soak items in this mix for an hour, or overnight if necessary, and rinse thoroughly in the morning.

Fine china
Restore fine china to its former glory with borax. Put on rubber gloves and dissolve 125 g (4½ oz) of borax in a sink of warm water. Soak your fine china for an hour and then rub at any stains with a soft cloth. Rinse thoroughly with cold water and dry.

Microwave cleaners

What are microwave cleaners?

Most people use all-purpose cleaners, antibacterial products or bleach to clean their microwaves. Read the chapters on these products to find out more about them.

Keeping it eco-friendly

Include your microwave oven in your daily cleaning routine. When the microwave has cooled down after use, remove any food spills with a soft cloth. Then, with a 50/50 cold water and white vinegar spray, spray around the inside and outside of your microwave and clean with a microfibre cloth. This daily maintenance will virtually remove the need for having to carry out a major cleaning job.

If you have a big spillage, then it is easier to remove and clean stains on the same day, rather than leaving them to dry. Allow things to cool, but try to clean them up as soon as you safely can. Use baking soda to clean off the remaining residue of any cooking disasters.

Homemade recipes

Microwave cleaner

Like conventional ovens, microwave ovens can become splattered with grease and food bits during cooking. Odours given off by grease and old food can impair the taste of the food you are cooking.

To clean a microwave oven, ensure all parts are cool before handling. Remove the glass turntable and wash this in a bowl of warm soapy water made with eco-friendly dishwashing liquid. To clean around the inside of the oven, use baking soda on a damp sponge. Baking soda has a gentle cleaning action that will neutralize odours and remove grease. Rinse well with warm water and spray with a 50/50 mix of water and white vinegar to remove any powdery residue. Repeat the process on the outside of the microwave. When you have cleaned and dried the glass turntable, place it back inside the microwave. Ensure the microwave is completely dry before using again.

Oven cleaners

What are oven cleaners?
Oven cleaners are available as aerosol cans, pump sprays or disposable wipes. They remove burnt-on grease, food and spills.

How do oven cleaners work?
Oven cleaners work by dissolving grease and grime. They literally 'eat away' hardened food deposits and oil so that you can rinse everything away and enjoy a spotless oven.

What's in oven cleaners?
Oven cleaners contain powerful corrosive agents and solvents to dissolve grease and strong perfumes. Cleaners in aerosol cans contain propellants and most wipes contain a large percentage of ethanol.

Dangers to you
Oven cleaners, along with drain cleaners, are two of the most toxic household products. Water and electricity are dangerous together, so as a precaution always turn electric ovens and hobs off at the mains before cleaning.

Powerful corrosive agents, such as sodium hydroxide (lye), are found in some conventional oven cleaners. A single exposure to sodium hydroxide can severely burn the skin or damage the eyes. Oven cleaners also contain a large proportion of solvents to dissolve grease or oil. The majority of solvents are irritants to eye, skin and mucous membranes. In addition, they can damage the lungs, kidneys and neurological system. Solvent exposure occurs during inhalation or through the skin.

Oven cleaners in aerosol cans are even more dangerous because they produce a fine mist of choking fumes that can be inhaled. This makes them particularly bad for people who suffer from asthma. This fine mist can also travel around your home and land on other surfaces such as your kitchen table or floor where children and pets play. Never use a conventional oven cleaner without good ventilation and wearing rubber gloves and a mask.

Dangers to the environment

Some solvents have been classified as toxic to aquatic life. According to Department for Environment Food and Rural Affairs (DEFRA), 1 litre (1¾ pt) of solvent can pollute over ten million litres of groundwater. Once solvents have entered the aquatic environment, they are extremely difficult to clean up. Some solvents dissolve rapidly in water and others sink, but both types cause pollution which can travel significant distances from the initial source.

Keeping it eco-friendly

Products that are so toxic should not really be used near you or surfaces you prepare food on. Cleaning the oven is probably the least liked cleaning task, so the following tips will mean you never need to scrub at it again!

- When cooking, keep food well covered, especially if it is oily. For example, if you roast a chicken, which has a tendency to spit fat as it cooks, invest in a roasting tin with a lid. The meat will still brown, even if it is covered. If you're cooking something that tends to bubble over, such as lasagna, then put the dish on an ovenproof tray to catch spills.
- After each use, wait for the oven to cool down and clean with either some baking soda, or wipe around with a 50/50 white vinegar and cold water mix to freshen and clean.
- After you've done a really thorough clean of your oven, line the bottom of the oven with aluminium foil to catch grease spills. Ensure the foil does not touch or obstruct the elements, fan assemblies, or vents in a gas oven.
- If you notice a greasy food spill or splatter in the oven or on the hob, then cover it with salt or baking soda while still warm and once it is cooled, you can clean it. The salt or baking soda will have already absorbed a lot of the grease, making your cleaning job easier.
- Keep your eye on pans that are coming to the boil so they do not boil over.

Homemade recipes

Simple hob cleaner

When the hob is cool enough to touch, dip a damp sponge or soft scourer in a jar of baking soda and use to remove any bits of food or greasy residue from the hob. Rinse well with warm water and follow with a 50/50 spray of cold water and white vinegar to remove any powdery residue.

Simple oven cleaner

This cleaner is simple, yet so effective and has to be tried to be believed! Lightly spray your oven with cold water. Sprinkle a layer of baking soda over the oven floor. Lightly spray with water again and leave to dry (or you can leave overnight). Spray with water again and use a sponge to wipe away the baking soda, along with all the grease and grime. For tough areas, use a palette knife, very fine steel wool or carefully use a razor blade. Rinse well with hot water or spray a 50/50 cold water and white vinegar solution to clean away the powdery residue.

Soda crystals – not for use on aluminium

Soda crystals are great for dissolving grease, so if you have a particularly tough cleaning job to do, then use this recipe instead. Put on a pair of rubber gloves. Make up a strong solution of soda crystals by dissolving 250 g (9 oz) of washing soda in 600 ml (1 pt) of warm water. Using a sponge or soft scourer, use this solution in your oven or on the hob. Rinse well with warm water. If you have any aluminium parts on your oven or have an aluminium hob, then do not use washing soda. Try one of the recipes mentioned above instead.

Steam cleaning

Steam cleaning is a chemical-free way to clean your oven and hob. Follow the manufacturer's guidelines and you'll soon enjoy a sparkling oven!

Self cleaning

Don't forget that your oven liners may be self cleaning. Usually a high temperature will simply burn off the residue. Read the instruction manual to check the right cleaning method for your appliance.

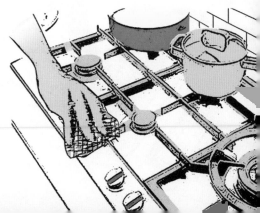

Sink cleaners

What are sink cleaners?
There are a number of specific sink cleaning products available, including liquids, gels and powders, but most people use all-purpose cleaners, antibacterial sprays or bleach.

How do sink cleaners work?
These products break down grease and oil, clean stains and help you to rinse dirt away.

What's in sink cleaners?
A typical sink cleaning and deodorizing product contains high levels of phosphates, some kind of bleach and perfumes, along with colourant.

Dangers to you
Sodium percarbonate, used as a bleaching agent in some conventional sink cleaning products, is irritating to the eyes and skin. Overexposure by inhalation may cause respiratory irritation. The label on one product reads 'Take care not to inhale product dust. Harmful if swallowed. Avoid contact with skin and eyes'. It also advises wearing rubber gloves, eye and face protection.

Dangers to the environment
Phosphates are particularly harmful for the environment. When phosphates enter rivers and lakes they promote excessive algae growth, known as algal bloom. This starves the water of oxygen and prevents sunlight from entering the water. The result is that fish die and the delicate ecosystem between plant and animal life is disrupted.

Keeping it eco-friendly
A few minutes spent wiping around sinks daily means they never again need to get greasy, dirty or with a ring of soap scum around them. Keep a spray

bottle containing a 50/50 mix of white vinegar and water next to the kitchen sink or washbasin. After you have brushed your teeth, washed your hands or done the washing up, wipe around the sink with this vinegar solution and a soft cloth. Using vinegar helps to prevent limescale build-up and will cut through soap scum.

Homemade recipes
Simple sink cleaner
The simplest solution is to use neat baking soda. Sprinkle some baking soda onto a damp sponge or cloth and use to scrub your sinks. Rinse well with warm water to remove any powdery residue.

Foaming sink cleaner
If you prefer a product that foams, put 1 tablespoon of baking soda in a bowl and mix in enough liquid castile soap to achieve a foaming consistency. Add 2 drops of lemon or orange essential oil if you wish. Use this like a cream cleaner for cleaning sinks, draining boards and around taps. Rinse well with warm water.

Sink cleaner for greasy areas
If you've washed up after a particularly greasy meal, you might need something stronger than baking soda. Put on rubber gloves and combine 1 tablespoon of soda crystals with 600 ml (1 pt) of warm water. Use this solution to dissolve grease. Rinse well with warm water.

Removing limescale and watermarks
Spray neat white vinegar onto draining boards or around taps – wherever you get limescale build-up. Leave for a few minutes and then wipe with a sponge or microfibre cloth. This is especially effective on stainless steel draining boards. Do not use neat vinegar on metal-plated finishes or fixings.

Scouring powder
Put on rubber gloves and mix together equal quantities of salt, baking soda and borax in a wide-mouthed jar with a lid. Use this scouring powder with a damp sponge or cloth to clean your sink or washbasin. It can even be used on plastic surfaces without scratching. Rinse well after use.

Dishwashing detergents

What are dishwashing detergents?
Dishwashing detergents are designed to
remove oil and grease from crockery,
glassware, cutlery and cookware.

How do dishwashing detergents work?
Surfactants in dishwashing detergents allow oil and grease to be emulsified
and dissolved in water. In addition, these products keep oil and grease
suspended in the water so your crockery comes out clean.

What's in dishwashing detergents?
Dishwashing liquids are made up of detergents, surfactants, preservatives,
colours and fragrances. Sodium laureth sulfate (SLS) is the main ingredient
in most dishwashing products.

Dangers to you
Most dishwashing detergents are highly fragranced, which can be irritating for
sensitive individuals. Symptoms range from watery eyes to coughing, asthma
and headaches. Dishwashing products are designed to pull oil and grease from
surfaces, which is why you often end up with dry hands after using them.
The same ingredients are pulling natural oils from your skin and drying them
out, which can lead to irritation. Make sure you rinse all traces of the product
off anything you will be eating off so that you do not ingest any chemicals.

Keeping it eco-friendly
Use a washing up bowl of water instead of a running tap when you wash
dishes to save water. If you are using a conventional dishwashing detergent,
then only use a tiny amount. We all think that foam is what gets things clean,
but it isn't. The more foam you have, the more rinsing you need and the more
water you use.

When I'm preparing a meal I have a bowl of water on the side to place
kitchen items in as soon as I've used them. This prevents food drying and
becoming difficult to remove.

Scrape any leftover food into the bin, compost caddy or wormery so that it does not contaminate the washing water or block the sink. When you have a full load of items to wash up, begin with the cleanest items first – usually glasses and cups. Move onto cutlery and then the crockery. Save the greasiest and dirtiest items, like pans and roasting trays, until last. This keeps the water cleaner for longer and means you do not transfer grease from one item to another.

Homemade recipes
Simple and effective
Most of us think we need lots of sweet smelling bubbles to get things clean, but this is not always true. For all but the greasiest items, or those that have had raw meat on them, hot water can be enough. If you wash up as you go along so that food does not dry onto plates and pans, you'll find that plain hot water is effective for many dishwashing jobs.

Dishwashing liquid
For foam lovers, fill a 500 ml (18 fl oz) squirty bottle with liquid castile soap and add 10 drops of essential oil of your choice. Lemon and tea tree are antibacterial.

Dishwashing without bubbles!
Dishwashing liquids cut through grease so that you can rinse dirt and residue away. Lemons and baking soda can do this too. Dissolve 1 tablespoon of baking soda in a bowl of hot water and mix in the juice of half a lemon. There will be no bubbles, but your dishes will be clean and you'll use less rinsing water.

Borax to cut through grease
Wear rubber gloves and dissolve 1 tablespoon of borax in a bowl of hot water. This will cut through grease effectively and is great for particularly dirty items. Rinse well after use.

Cleaning burnt saucepans
A burnt saucepan is one of the worst items to have to clean! Fortunately help is at hand in the form of baking soda. Sprinkle a thick layer of baking soda over the burnt food. Leave on overnight and in the morning you'll be able to easily scrape away the burnt bits of food and clean your saucepan.

The laundry room

Most people have their favourite laundry products – usually determined by smell and how effective they are. But did you know that the reason you get 'whiter whites' or 'brighter brights' is usually due to chemical residue being left on your fabrics? In this chapter you'll find out how to make your own simple products that do not leave toxic residues on your skin or use scents that make your eyes water!

Fabric softeners and dryer sheets
What are fabric softeners and dryer sheets?
These products soften fabrics, reduce static cling and add fragrance to linens. Some products claim to make ironing easier or clothes dry faster.

How do fabric softeners and dryer sheets work?
Fabric softeners work by depositing lubricating chemicals on fabrics, making them feel softer and reducing static cling. Dryer sheets are coated with chemicals that transfer onto your clothing when the dryer heats up.

What's in fabric softeners and dryer sheets?
Fabric softeners contain synthetic fragrances, surfactants and preservatives. Dryer sheets contain softening agents, antistatic agents and perfumes.

Dangers to you
Chemical residues from fabric softeners and dryer sheets are against your skin all the time you wear clothes or sleep on bedding that has been washed with these products. Consequently, they are not recommended for baby's clothes. Dryer sheets eliminate fire retardant treatment on clothing, so should not be used on night clothes. Both these products are usually highly scented, so people with sensitivities to artificial fragrances should be particularly careful about which products they use. Some fabric softeners contain formaldehyde which is a suspected human carcinogen. It can cause itchy, watery eyes, stuffy nose and breathing difficulties.

Dangers to the environment
Fabric softeners often come in large plastic containers which end up in the landfill. Dryer sheets are once-use disposable items that end up in the landfill.

Keeping it eco-friendly
• Line dry your clothes rather than use a tumble dryer, wherever possible.
• If you do use a dryer, save electricity and get less static by taking your clothes out of the dryer while they are still slightly damp.
• Use a gentle spin on your washing machine. The faster your clothes spin, the more the fibres get 'squashed', making your clothes feel rougher.
• If you do use a dryer, don't overload it as your clothes will take longer to dry.

Homemade recipes

Easy fragrant dryer sheets
Find a piece of material such as a small muslin square or handkerchief and add 4 drops of your favourite essential oil to the sheet. Place the sheet in the dryer with your clothes. The action of tumble drying on its own will soften your clothing and towels, especially if they are made from natural fibres. There is then no need for chemical softeners.

Chemical-free drying
Try dryer balls. As well as reducing static, they soften clothes, last for years and do not contain chemicals. Dryer balls can be found in health food stores and 'green shops'.

Simple fabric softener
Add 250 ml (9 fl oz) of white vinegar to the rinse cycle. Vinegar helps to remove soap and detergent residue, and prevents static cling in the dryer.

Soda crystals
Soda crystals have long been used as a water softener. Add 100 g (3½ oz) to your wash load for soft, fluffy clothes. If you are in a hard water area, adding washing soda to your laundry means you can then put in the recommended amount of laundry detergents for 'soft water' areas. It also means you can forgo using harsh fabric softeners.

Baking soda
Add 50 g (2 oz) of baking soda to your wash cycle to help soften clothes.

Magic fabric softener
Mix together 450 ml (16 fl oz) of white vinegar with 250 g (9 oz) of bicarbonate of soda and 850 ml (1½ pt) of cold water. The bicarbonate of soda and vinegar will fizz, so combine them slowly and carefully over the sink. Add 4 drops of your favourite essential oil to this if you wish. Pour into a storage bottle, cover and shake, releasing any gas build-up frequently. Use 60 ml (2 fl oz) in the final rinse by pouring the mix into the fabric softener compartment of your washing machine.

Laundry detergents

What are laundry detergents?
Laundry detergents are liquids, powders and tablets that remove stains from clothes, brighten colours and whiten whites. They make fabrics smell and look clean.

How do laundry detergents work?
Ingredients in laundry detergents help soften water, lift stains from fabrics, remove odours, prevent dirt from settling back on fabrics, prevent corrosion to your washing machine and scent fabrics.

What's in laundry detergents?
Surfactants are the 'work horse' of laundry products. When dissolved in water, surfactants enable dirt to be loosened and removed from surfaces. They also keep dirt suspended in the water so that it can't settle back onto the fabric being cleaned.

Laundry detergents contain optical brighteners. They coat the material with fluorescent particles which reflect light. This gives the appearance of bright

colours and white clothing. Bleaches are used to make whites look whiter and to take the colour out of stains to make clothes look cleaner. Perfumes and colours do not have a cleaning function, they just make the product more attractive to consumers and scent your clothes. Biological products contain enzymes. These literally 'eat' stains on fabrics.

Dangers to you
You only need to soak a recently washed, dry cloth in a bowl of clean hot water, then wring out well to see

the white foamy residue come out of it! We know through the use of
hormone and nicotine patches that the skin is porous, so maybe some of the
chemicals in laundry products can enter the bloodstream. In fact, it has been
found that up to 60 per cent of some chemicals enter the skin solely from
surface contact.

You are potentially in contact with laundry detergent all day, every day.
When you go to bed you are sleeping on linen that has been washed in
your chosen product and when you step out of the bath or shower and wrap
yourself in a towel, you are wrapping yourself up in freshly laundered fabrics.
Babies chew everything, including their clothes, which puts them at higher risk
from ingesting the ingredients in your laundry products.

Bleach
Bleaches such as sodium hypochlorite, can cause skin irritation and eye
damage. Studies have shown that sodium hypochlorite and organic chemicals
such as surfactants and fragrances contained in several household cleaning
products react to generate chlorinated volatile organic compounds (VOCs).
Most of them are toxic and probable human carcinogens.

Optical brighteners and enzymes
Some optical brighteners can cause allergic reactions when in contact
with skin. Products with enzymes can irritate the skin, causing rashes and
dermatitis. This is why non-biological products are recommended for babies
and people with eczema.

Surfactants
Linear alkylbenzene sulfonate (LAS) is the most widely used surfactant in the
world. Linear alkylbenzene (LAB), the material used to produce LAS, is derived
exclusively from petroleum derivatives: benzene and linear paraffins. Breathing
in low levels of benzene can cause drowsiness, dizziness, rapid heart rate,
headaches, tremors, confusion and unconsciousness. Benzene damages the
bone marrow and can cause a decrease in red blood cells, leading to anaemia.
Benzene causes leukemia and is associated with other blood cancers. It is a
known carcinogen.

Dangers to the environment

Most conventional cleaning products use petrochemical-based ingredients derived from fossil fuels which are non-renewable. Residues from these are poorly biodegradable and build up in the environment where they cause significant, long-lasting damage to humans, animals and the environment. Phosphates are a major hazard to the environment and can cause irreversible damage to the aquatic environment by removing oxygen from water.

Keeping it eco-friendly

- There are a number of effective, natural laundry detergents on the market. Look for ones made with plant ingredients that readily biodegrade and contain no phosphates.
- Use a cold or 30-degree wash wherever possible.
- Only do a machine wash when you have a full load and use the economy setting to make full use of any detergent, water and electricity.
- Try reducing the amount of detergent you use. Often, you can get away with using half the amount recommended by manufacturers.
- Softening the water means you can use less detergent. Add 100 g (3½ oz) of soda crystals to your machine and try using less detergent.
- If you spill something on clothing or linens, deal with it immediately to prevent the stain drying and becoming more difficult to remove.

Homemade recipes

As always, never mix homemade recipes with commercial products as unknown and possibly unwanted effects may occur.

Pure soap

If you're lucky enough to have soft water, you can use pure soap flakes. If you try to use soap in a hard water area, you'll end up with an insoluble film that can turn clothes dull. Dissolve 25 g (1 oz) of grated castile soap in hot water and put in the wash cycle (either directly in the washing machine or in the drawer). You'll need to use a cold water rinse if using soap and will need to experiment with amounts to get it right.

Borax and vinegar

Put on rubber gloves when handling borax. Add 50–85 g (2–3 oz) of borax to the wash cycle and 250 ml (9 fl oz) of white vinegar to the rinse cycle (in the fabric softener compartment). To boost stain removal, add 250 ml (9 fl oz) of 3 per cent hydrogen peroxide to the wash cycle. 115 ml (4 fl oz) fresh lemon juice added to the final rinse cycle will brighten clothes.

Laundry gloop

This is a popular alternative to conventional detergent. Put 4 litres (7 pt) of water in a large pan and bring to the boil. Meanwhile grate 140 g (5 oz) of unscented pure soap (castile is good). When the water has boiled, add the soap and turn down the heat. Stir until all the soap has dissolved. Remove from the heat and add 125 g (4½ oz) of soda crystals, stirring until they have dissolved. Allow to cool, then decant into a large enough tub or a number of smaller ones. Use about half a teacup for each load.

Note: to avoid potential problems with machine blockages, put the gloop directly into the drum, not the drawer. Add 115 ml (4 fl oz) of distilled white vinegar to the final rinse to reduce any soap build-up.

Gloop alternative

Put on rubber gloves and mix together 140 g (5 oz) of grated pure soap, such as castile, with 125 g (4½ oz) of soda crystals and 125 g (4½ oz) borax. Use 1 tablespoon for a light load. For heavily soiled loads, use 2 tablespoons.

Chemical-free laundry

For a chemical-free wash, why not try soapnuts or Ecoballs? Both are available from health food stores and some 'eco stores'. Soapnuts come from the Sapindus tree and contain saponin, a natural soap that forms gentle bubbles when the soapnuts come into contact with water. Ecoballs are a set of three 'balls' that you put in the washing machine, without the need for detergent. They ionize the water, allowing it to get deep into your clothes and lift the dirt away. Ecoballs are reusable for up to 1,000 washes.

Stain removers, pre-soaks and whiteners

What are stain removers, pre-soaks and whiteners?
These products are liquids, solid sticks, sprays and powders, used to remove stains from fabrics and whiten whites. You can buy general brighteners and whiteners, pre-soaks or specific spot removers such as ones for ink, grass, mud, oil and blood.

How do stain removers, pre-soaks and whiteners work?
They work by bleaching colour from stains or by lifting stains from fabrics so that they can be washed away.

What's in stain removers, pre-soaks and whiteners?
These products have similar ingredients to laundry detergents and contain bleaching agents, surfactants, enzymes, optical brighteners, perfumes and colourings.

Dangers to you
Some products contain sodium carbonate peroxyhydrate which is harmful if swallowed and irritating to the skin. It causes risk of serious damage to eyes. Some products with this ingredient recommend you wear eye and face protection as well as rubber gloves. Sodium lauryl sulfate (SLS) pulls oils and fats from surfaces, so can be highly irritating to the skin.

Artificial fragrances and colourants can be extremely irritating. Some people report sneezing fits, watery eyes, headaches and respiratory difficulties after coming into contact with artificial fragrances. Bleaches are irritating to skin, eyes and lungs, and are corrosive and harmful if swallowed.

Dangers to the environment
You'll find warnings such as 'Dispose of this material and its container at hazardous or special waste collection points' on some stain removers. This means they contain dangerous ingredients that shouldn't get into the watercourse or in the landfill. Better to avoid them altogether, don't you think?

Keeping it eco-friendly
Prevention is better than cure, so the best thing to do is to treat stains as
soon as possible. This will result in less water, less cleaning product, less
energy and less waste. It will also save you time and frustration!

Stain busters
Borax, salt and baking soda: these products are helpful for absorbing
grease. Sprinkle them onto a stain and leave to work for an hour or
longer – overnight is good!

Club soda: club soda helps to 'fizz out' tough stains such as red wine, coffee
and blood, and bring them to the surface so that you can rinse them away.
Pour liberally on stains.

Detergent and soap: pre-treating stains by rubbing in environmentally-
friendly detergent or pure soap such as castile, can be as effective as a
chemically-laden product. Adding a little vegetable glycerin to the detergent
can make it more effective (glycerin is available from health food shops and
some 'green' shops). Mix together a tablespoon of vegetable glycerin with a
tablespoon of environmentally-friendly dishwashing
detergent and add 6 tablespoons of water. Apply to
stains with your fingertips. This formula works
well on ink, coffee, tea and grease stains.

Hydrogen peroxide: add 115 ml (4 fl oz) of
3 per cent H_2O_2 to the rinse cycle of your
machine for general whitening, or use as a
pre-soak for stains such as blood. Do not use
hydrogen peroxide if you are using conventional detergents in the main wash.

Vinegar: is effective as a pre-soak for sweaty
clothes such as gym kits. Pour neat white vinegar
into a spray bottle and spray onto the sweaty parts
of clothing, such as under the arms or on socks.

Alternatively, pour 250 ml (9 fl oz) of white vinegar in to the pre-wash and let the fabrics sit for an hour before laundering.

Natural whiteners: sunlight is a free and effective bleaching agent! Line dry your whites on a windy, sunny day and let nature do the work. If the sun is not shining, then lemon juice has a mild bleaching effect. Add 115 ml (4 fl oz) of fresh lemon juice to the rinse cycle of your machine. Other ways to whiten are to add either 125 g (4½ oz) of borax or soda crystals to a load of laundry.

Homemade recipes
Getting rid of tough stains
There are many books available on specific stain removal. Here are a few ideas to get you started.

Blood and egg: soak the garment in cold salt water. Leave overnight and launder the following day. Alternatively, soak in a strong solution of soda crystals. Put on rubber gloves and add 250 g (9 oz) of soda crystals to 600 ml (1 pt) of water. Soak the garment for an hour, rinse and wash as usual. If the stain is particularly stubborn, soak the stain in 3 per cent hydrogen peroxide before laundering.

Grass: soak the garment in a 50/50 mix of cold water and white vinegar overnight. Rinse and wash the following day.

Grease: butter, cooking oil, chocolate or salad dressing can be removed with a dry powder of equal quantities of salt and baking soda. Leave the powder on the stain for an hour. Brush away the excess and launder. Alternatively, soak in a strong solution of soda crystals, as above. Soak the garment for an hour, or overnight if it is really greasy, then rinse and wash as usual.

Ketchup, barbecue sauce and tomato-based stains: soak in a 50/50 solution of cold water and white vinegar for an hour and wash as usual, adding 250 g (9 oz) of borax to the washing machine powder drawer.

Washing machine maintenance

No matter how well you look after it, sometimes a washing machine can start to smell musty. It is far better for the environment and your bank balance to take good care of your machine with the following tips.

Check the pipes: is there a kink or bend in the pipe where water can stagnate? Straighten pipes as much as you can and give them space to move when the water flows through them.

Check the filters: washing machine filters are designed to trap anything that shouldn't go down the drain such as hair, dust and fluff. If the filters become blocked, the machine cannot rinse and drain properly. Check the filters once a week and keep them clean.

Reduce your detergent: you could be using too much detergent which can result in a build-up of product residue in the machine and on your clothing. Experiment with reducing the amount of laundry product you use.

Take the washing out straight away: leaving your washing in the machine once it has finished laundering can lead to musty smells. Take the washing out and hang to dry as soon as possible after the cycle finishes.

Leave the door open: leave the door of the machine slightly open between washes. Remember to check for pets, toys or food before doing your next wash if you have curious toddlers in the house!

Do a hot wash: doing a hot wash uses a lot of resources, but the occasional hot wash will clean the pipes of your machine thoroughly. Pour 250 ml (9 fl oz) of white vinegar into the soap dispenser and run the machine on the hottest cycle. This will cut through any soap scum and limescale in the pipes. To assuage your guilt, fill the machine with white bedding sheets, nappies or towels!

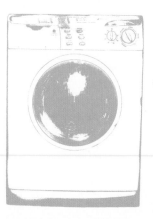

Shoe polishes

What are shoe polishes?
Shoe polishes add shine and protect shoes.
They come as solid waxes which are applied
with shoe brushes or rags, creams which come with a
brush, liquids which come with sponge applicators and disposable wipes.

How do shoe polishes work?
Shoe polishes apply a protective wax coating to shoes that repels water
and stains. They add shine and restore the appearance of leather footwear.

What's in shoe polishes?
Shoe polish is usually a waxy emulsion, made from a mix of natural and
synthetic materials including naphtha, lanolin, turpentine, wax, ethylene glycol,
thickeners and dyes. Shoe polishes need to contain a high level of volatile
substances so the polish dries and hardens after application.

Dangers to you
Shoe polish can be toxic, and stain your skin and clothing. It contains
chemicals which can be absorbed through the skin so you should wear gloves
when handling conventional shoe polish. It will irritate the eyes if you are
unfortunate enough to get some in your eyes. Spray shoe polishes and
protectors mean the product is easily inhaled. These products should be
used in well-ventilated areas.

Many shoe polishes contain solvents such as nitrobenzene, methylene chloride
and naptha. Nitrobenzene is highly toxic and readily absorbed through the
skin. Prolonged exposure may cause serious damage to the central nervous
system, impair vision, cause liver or kidney damage, anemia and lung irritation.
Inhalation of fumes may induce headaches, nausea, fatigue and dizziness.
Methylene chloride is a suspected carcinogen, linked to cancer of the lungs,
liver and pancreas in laboratory animals. Its high volatility makes it an acute
inhalation hazard. Methylene chloride is metabolized by the body to carbon
monoxide potentially leading to carbon monoxide poisoning. Pregnant women
must avoid contact with this product because it crosses the placenta.

Dangers to the environment
Some of the ingredients in shoe care products are obtained from oil – a non-renewable resource – so are best avoided.

Keeping it eco-friendly
Waterproof new leather shoes before wearing them by using the recipe for shoe salve (below). Apply every month for long-lasting protection. If your shoes get wet, never dry them near a direct source of heat as this can dry out and crack the leather. Stuff shoes with newspaper to retain their shape. Once they are completely dry, reapply the waterproofing before wearing them again.

Homemade recipes
Footwear must be clean and dry before applying shoe polish. Polishes are not cleaning agents; they are finishing and protecting agents to restore shine and appearance of your shoes. The main part of the shining is done by buffing and a microfibre cloth is enough to achieve this for everyday maintenance.

Shoe salve
This simple salve provides a protective, waterproof layer for your footwear. Gently heat 25 g (1 oz) of beeswax and 115 ml (4 fl oz) of cheap olive oil in a double boiler until they melt, stirring all the time. When melted, remove from the heat and stir until cooled. Store in a glass jar with a tight-fitting lid. Use with a soft cloth to apply a layer of polish to your shoes.

Simple shoe treatment
Apply a cold-pressed oil such as inexpensive olive oil to your footwear, then buff well with a microfibre cloth until they shine.

Bananas!
Use the outside of a banana peel as an innovative and 'no-waste' way to add shine to your shoes!

Spit
If you have nothing in the house to clean your shoes with, then you can use one of your own natural products: spit! Simply spit a few drops onto your shoes and buff with a microfibre cloth.

The bathroom

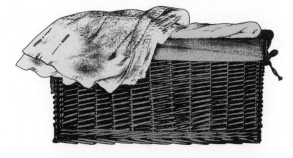

Your bathroom can be a haven in which to escape at the end of the day. It can also be a germ-ridden nightmare that haunts your sleep! There are many harsh products designed to kill germs, remove limescale and leave your toilet sparkling, but many contain ingredients that are best avoided. Here you will find some recipes to keep things fresh and clean, so you can enjoy a bath or shower without concern about bathing in a toxic soup.

Bathtub cleaners

What are bathtub cleaners?

Bathtub cleaners are sprays, liquids, creams, powders and wipes designed to tackle limescale, remove soap scum and help prevent a build-up of both.

How do bathtub cleaners work?

Bathtub cleaners are similar to all-purpose cleaners with added ingredients to tackle limescale and prevent soap scum build-up. Many bathtub cleaners contain antibacterial ingredients too.

What's in bathtub cleaners?

A typical product for cleaning baths contains surfactants, ethylenediaminetetraacetic acid (EDTA), ammonium chloride, perfumes, solvents and preservatives.

Dangers to you

EDTA has been found to be both cytotoxic (toxic to cells) and weakly genotoxic in laboratory animals (genotoxic substances are known to be potentially mutagenic or carcinogenic). Butylphenyl methylpropional is a synthetic fragrance. It is a skin irritant and in tests done on animals, high concentrations applied to the skin caused damage to sperm and the central nervous system. Most conventional products contain warnings on the labels not to breathe in the spray, to avoid contact with skin and eyes, and to use only in well-ventilated areas but it is difficult not to breathe in the spray when you are bent over a bathtub in a small space like the average bathroom.

Dangers to the environment

EDTA collects in groundwater and is an environmental concern because it has strong metal-chelating properties. This has the potential to remobilize heavy metals from river sediments. In turn, these metals can get carried back into our drinking water supplies and food, especially fish.

Keeping it eco-friendly

Prevention is definitely better than cure in the bathroom, otherwise it can become a breeding ground for mildew because it is such an enclosed damp space. Bacteria thrive in warm, damp places, so ventilate your bathroom well. Open a window or turn on an extractor fan when you run a hot bath or shower. When you have finished using the bathroom, leave a window open, the fan running and the door open to allow steam and vapours to escape. After a bath, wipe around the bath with a sponge dipped in your chosen cleaner. Give it a quick spray with a 50/50 mix of cold water and white vinegar and you'll never have a dirty tub to clean again. It makes your next bath so much more pleasurable knowing that the bath is clean at any time for you to enjoy!

Homemade recipes

All purpose-cleaners

You can use any of the cleaners recommended in the all-purpose cleaners section (see pages 45–7). My favourite is the cream cleaner made from baking soda, liquid castile soap and essential oils (see page 47). Why not choose some decadent essential oils, such as rose and benzoin for a relaxing and uplifting scent?

Baking soda

Sprinkle some baking soda onto a damp sponge and scrub away. Baking soda will remove deposits, and after rinsing well with warm water, you'll be rewarded with a sparkling, clean tub. Finish with a 50/50 mix of cold water and white vinegar spray to clear any powdery residue and prevent limescale if you live in a hard water area.

Soda crystals

If you inherit a greasy tub, then soda crystals will help clean it but do not use soda crystals on lacquered taps and fittings. Put on rubber gloves and make a solution of 1 tablespoon of soda crystals to 600 ml (1 pt) cold water. Use this solution with a sponge to dissolve grease. Rinse well with warm water.

Natural scouring powder

Again, if drastic action is needed on your bathtub, then the following, non-scratching scouring powder can be used. Put on rubber gloves and mix together equal quantities of salt, baking soda and borax. Sprinkle onto a damp sponge or cloth and watch the soap scum and grease dissolve away as you scrub. Rinse well with plenty of water.

Mould and mildew cleaners

What are mould and mildew cleaners?

Mould and mildew cleaners are liquids or sprays designed to kill mould spores and prevent the spread of mould and mildew. Mould and mildew can appear anywhere but in this chapter we will deal with the bathroom, as this is one of the most common places.

How do mould and mildew cleaners work?

By using strong bleaches, mould and mildew cleaners kill mould spores and prevent regrowth.

What's in mould and mildew cleaners?

Mould and mildew products contain bleaches, surfactants and perfumes. Bleach is the active ingredient, and as such, these products usually contain a high percentage of bleach.

Dangers to you

Unfortunately, mould itself can be detrimental to health, so you do need to deal with it. Mould spores can be allergenic, causing irritation to eyes, nose, throat and lungs. Sodium hypochlorite (bleach), when combined with other ingredients such as surfactants and fragrances, can create chlorinated volatile organic compounds (VOCs). Most of these are toxic and suspected human carcinogens. Sodium hypochlorite is corrosive and can cause burns to the skin. Many people suffer from irritated eyes and throat when using these products.

Warnings on typical mould and mildew products include 'Do not breathe spray mist' and 'Do not contaminate foodstuffs, eating utensils or food contact surfaces'. It is very difficult to avoid breathing in the spray mist when working in confined areas such as a bathroom or shower unit, so it is vital that you work in a well-ventilated area and use face and skin protection.

Dangers to the environment

The label on one product reads 'This product contains substances which are known to be hazardous to the environment. Do not contaminate ground waterbodies or watercourses with chemicals or used container'. Conventional mould and mildew cleaners should not be used near drains and surface run-off should not enter drains. If you are told to leave a product to work then rinse thoroughly and wipe off, where does that product run-off go to? Surely it will end up down the drain... It is also important to remove or cover all fish tanks or bowls and avoid contact with plant life. It's time for a safer alternative, don't you think?

Keeping it eco-friendly

Most mould and mildew cleaners are only a short-term solution, so it is far safer to make your own. Removing mould promptly is important because it can spread quickly. Mildew is a living fungus that can have harmful effects on your health. Mould thrives in moist, dark, warm areas with no ventilation. Take away one of those conditions and the mould will have more trouble thriving. A bathroom is the prime place for mould and mildew to proliferate, but there are steps you can take to prevent mould becoming a problem.

Starve it! The primary source of 'food' for mould is damp. Ensure there is good ventilation in your bathroom. If it has windows, open them, especially when running a bath or shower. If there are no windows, then fit and use an extractor fan. Finally, when you leave the bathroom, dry the floor, take out damp towels and air them, and keep the door open to let air flow around the room.

Root cause. Look for the root cause of damp in your home. Is broken or leaking guttering contributing to damp walls? Does your home have damp proofing? Are there any leaking pipes?

Once every couple of months, take your shower curtain down. Launder, or take it out and air on an outside line on a sunny day. Keep a spray bottle with a 50/50 mix of water and white vinegar next to the shower. After your shower, spray this onto the tiles and leave. This will help prevent mould and mildew spreading.

Homemade recipes

Borax

Put on rubber gloves and dissolve some borax in a little hot water to make a paste. Apply the paste with an old toothbrush directly onto the mildew. Leave the solution on for an hour or a few days if the problem is severe. Dust away, rinse with warm water and dry thoroughly.

Soda crystals

If you have mildew or mould on your shower curtain, then use soda crystals. Put on rubber gloves and dissolve 250 g (9 oz) of soda crystals in 600 ml (1 pt) of hot water. Use this solution to wash your shower curtain. Try to line dry it in strong sunlight after this treatment for extra mould-killing power!

Vinegar

Vinegar can be helpful if the mould or mildew is not too severe. Mix equal parts of white vinegar and cold water in a spray bottle. Apply to the mould and leave to work. There is no need to rinse. For mouldy fabrics, such as a towel left stuffed in a swimming bag, pre-soak in a bucket or sink of cold water with 250 ml (9 fl oz) of neat white vinegar and 2 drops of tea tree essential oil. Soak for an hour before laundering as usual and drying in the sun.

Essential oils

Combine 25 drops of tea tree oil with 450 ml (16 fl oz) cold water and 3 tablespoons of liquid castile soap in a spray bottle. Use this foaming spray on mould and mildew. Spray on and leave to work for a few days. Brush away any mould and rinse well.

Grout and tile cleaners

What are grout and tile cleaners?

Tile and grout cleaners are alkaline cleaners, strippers and degreasers. They are designed to remove grease, soap scum, body oil and mildew. You apply them, leave to work for a few minutes and rinse off.

How do grout and tile cleaners work?

Powerful alkaline ingredients cut through grease and grime so you can rinse the dirt and build-up away. In addition, these products usually contain bleaches to take the colour out of stains and kill mildew.

What's in grout and tile cleaners?

The primary active ingredient is a powerful alkaline such as sodium hydroxide, along with preservatives, artificial fragrances and surfactants.

Dangers to you

Sodium hydroxide is an alkaline that in strong concentration can cause chemical burns, scarring and blindness. It is caustic and can dissolve grease and oils in the skin making it extremely damaging to humans. For this reason, it is advised on the labels of most tile and grout cleaners to use both rubber gloves and eye protection, and to avoid all skin contact.

Keeping it eco-friendly

Instead of allowing tiles and grout to get dirty and greasy, a daily or weekly wipe over with a 50/50 mixture of cold water and white vinegar will prevent limescale and soap scum build-up. Keep a spray with this mix next to the tub or shower unit so that it is there all the time. Apply after a bath or shower, or spray on in the morning as part of your home cleaning routine.

Homemade recipes

Tile and grout restorer

Spray neat, white vinegar over the surface of your tiles, leave for up to an hour before using a soft scrubbing brush. This will remove limescale, surface grease and restore shine.

Multi-purpose cleaner
You might recognize this recipe from the All-purpose cleaners section (see page 47). It's my favourite and I use it for most of my cleaning jobs. In a lidded container mix together 85 g (3 oz) of baking soda, 4 fl oz (120 ml) of liquid castile soap and 6 drops of your favourite essential oils (optional). You'll need to shake the mix before each use, but it works beautifully and brings a wonderful shine to surfaces. Rinse after use with warm water.

Grout 'scourer'
Cut a lemon in half and dip the cut surface in salt. Use this as a 'scourer' for the grout. Replenish the salt as necessary and rinse the grout after cleaning.

Essential oils
You can use neat tea tree to remove mould and mildew on grout, although it's not very economical. Dip an old toothbrush in the essential oil and scrub away. Tea tree can smell very strong, so use in a well-ventilated area or dilute it in cold water.

Soda crystals
Put on rubber gloves and mix 125 g (4½ oz) of soda crystals with 600 ml (1 pt) of water. Use this to clean tiles and grouting.

Steam
A steam cleaner will bring your grout up like new. Follow the manufacturer's guidelines for use.

Baking soda
Make a paste from baking soda and water. Apply this to grout with a soft brush. Leave to dry and rinse off. You can also use baking soda on tiles. Sprinkle some onto a damp sponge, wipe over the tiles and rinse.

Borax
Wear rubber gloves and mix together equal quantities of borax and baking soda. Put some on a moistened brush and use to scour tiles and grout. Leave on for an hour and rinse well with warm water.

Limescale removers

What are limescale removers?
Limescale removers, or 'descalers', remove limescale from areas like tiles, taps, bathtubs and toilets. They are available in liquids, sprays or solid blocks.

How do limescale removers work?
Bleach whitens limescale, but cannot remove it, so limescale removers contain strong acids to dissolve the limescale. Limescale blocks that sit in your cistern do not clean limescale; they prevent it forming, so you need to clean the limescale first.

What's in limescale removers?
The main acid used in powerful descalers is hydrochloric acid. In addition there are surfactants, perfume and disinfectants.

Dangers to you
Strong acids, found in many limescale removers, can cause severe burns and contact with skin and eyes must be avoided. A warning label on one brand advises the use of protective clothing, gloves, eye and face protection. Hydrochloric acid is highly corrosive and must be handled with appropriate safety precautions. It must never be mixed with bleach because it combines to produce toxic chlorine gas.

Keeping it eco-friendly
It's a challenge to apply the 'prevention is better than cure' rule here, because limescale is inevitable in hard water areas. However, getting rid of it before it builds up too much means much less work for you in the long term.

Homemade recipes

There is one marvellous kitchen ingredient for removing limescale. It's safe and even edible! It's white vinegar.

Metal showerheads

If you have a metal showerhead that is scaled, boil it for 15 minutes in a solution of 250 ml (9 fl oz) of white vinegar and 1 litre (1¾ pt) of water. The limescale will break up and you can scrub it with a soft brush to remove it all.

Plastic showerheads

If your showerhead is plastic, then do not boil it. Instead, soak in a 50/50 solution of vinegar and hot water. Leave to soak overnight and scrub off any remaining limescale in the morning with a soft brush.

Taps

For taps, soak an old cloth in a solution of 50/50 white vinegar and hot water. Place this cloth against the limescale on the taps and leave to work for an hour or so. Scrub with an old toothbrush to remove any last deposits.

Toilet bowls

Add around 250 ml (9 fl oz) of white vinegar to the toilet bowl and leave overnight before scrubbing with a toilet brush. If the toilet is really scaled up, you may need to do this more than once, but perseverance will pay off.

Shower cleaners

What are shower cleaners?
Shower cleaners are liquids, sprays, powders and disposable wipes that remove limescale and soap scum. They add sparkle and shine to shower doors and tiles.

How do shower cleaners work?
Acids in shower cleaners dissolve limescale and soap scum to leave surfaces clean and shiny.

What's in shower cleaners?
Most shower cleaners are little more than all-purpose cleaners, with some added ingredients, such as strong acids, to cut through limescale and soap scum. The main ingredients are acids, surfactants, preservatives and perfume.

Dangers to you
A heavily scented product will be more intense once the shower area is full of hot steam. Artificial fragrances can trigger asthma and allergic reactions. EDTA, found in some shower cleaners, can be irritating to the skin and can lead to dizziness, headaches, nausea, sneezing fits and asthma attacks.

Dangers to the environment
Most conventional shower cleaning products are based on petroleum, a non-renewable resource. Many wipes are a once-use disposable item made from a cellulose and polyester blend that ends up in the landfill. The discharge of products containing toxic ingredients into wastewater is a well-known problem, causing pollution of water resources and constituting an ecological risk for aquatic organisms. Some surfactants are known to be toxic to animals and the ecosystem, and can increase the diffusion of other environmental contaminants.

Keeping it eco-friendly
Keeping things clean in the first place reduces the need for harsh cleaning products and lots of rinsing water which is better for the environment, and your bank balance! Showers are more eco-friendly than baths. They use, on average, one-fifth of the water needed for a bath. Bear in mind that some power showers can use *more* water and energy than is needed for a full bath.

Homemade recipes
Safer acids
There are two safe acids readily available in your kitchen that will cut through limescale and soap scum: vinegar and lemons. In addition, these two ingredients are deodorizers, neutralizing and eliminating smells. If limescale build-up on glass shower doors or ceramic tiles is particularly bad, you can use these products neat. Spray on neat white vinegar or fresh lemon juice and leave to work for 30 minutes. Rinse with warm water. For everyday freshness, keep a spray bottle near the shower with a 50/50 mix of water and white vinegar. Use this regularly on tiles and the shower door. Spray on and leave; there is no need to rinse. For dealing with limescale on shower heads, refer to the section on limescale (see pages 101–02).

Baking soda
Baking soda is a safe, mild abrasive which can be used as a 'scouring powder'. Sprinkle some baking soda onto a damp sponge and wipe around tiles and glass doors. Rinse with warm water to remove residue. Follow with the 50/50 white vinegar/water spray mix if you wish.

Cream cleaner
If you prefer a cream cleaner consistency, it is easy to make your own. Choose your favourite essential oils to make this a luxuriously scented product – who said housework needs to be boring!

Mix together 85 g (3 oz) of baking soda with approximately 120 ml (4 fl oz) of liquid castile soap and 6 drops of your favourite essential oils (try geranium and rose for a delicious scent). Mix together in a lidded container. You'll need to shake it before each use, but it works beautifully and will bring a shine to your shower tiles and glass door.

Shower curtains
If you have a shower curtain, then wash it regularly to prevent the build-up of soap scum, limescale and mildew. Spray it with neat white vinegar if hard water deposits are a problem in your area before washing. See the section on Mould and mildew on pages 96–8 for further information.

Toilet cleaners

What are toilet cleaners?
Toilet cleaning products are liquids, gels, disposable wipes or solid blocks that kill germs, get rid of limescale, clean stains and get rid of odours in the toilet.

How do toilet cleaners work?
They kill germs with disinfectants, dissolve or prevent limescale with powerful acids, clean stains with bleaches and get rid of odours with the help of artificial fragrances.

What's in toilet cleaners?
Depending on the product, you will find acids, surfactants, perfume, disinfectant, colourings, bleaching agents and preservatives.

Fragrances and formaldehyde
Strong fragrances can cause headaches, watery eyes, sore throat and nausea. Artificial fragrances can trigger asthma. Some artificial fragrances are endocrine disruptors which can cause cancer. They are particularly bad in products that sit in the cistern or hang over the rim of the toilet, as every time you flush, a mist of product enters the air. Formaldehyde kills most bacteria and fungi and is a preservative. Formaldehyde can be toxic, allergenic and carcinogenic, and is a common indoor air pollutant. It can cause a burning sensation in the throat and breathing difficulties.

Dangers to you
Some toilet cleaning products are harsher than others, but none should get on to the skin or in the eyes. Those with strong acids can cause severe burns or serious damage to the eyes. Labels on most toilet cleaning products recommend the use of rubber gloves, face and eye protection. Many products contain highly corrosive ingredients which must not be swallowed.

Dangers to the environment
Due to their large-scale use, several types of synthetic fragrances, in particular musks, persist in the environment, accumulate in animals, and have been found

in human fat and milk. These pollutants pose additional health problems when they enter human and animal diets. Disposable wipes are often made from polyester, which is derived from petroleum, a non-renewable resource.

Keeping it eco-friendly

Not many people enjoy cleaning the toilet or using a dirty one, so preventative measures are best. A daily 'swish and swipe' as FlyLady calls it (see page 126) is all that is required. Pour your chosen homemade cleaner down the toilet every night or every morning, whatever suits your routine. Leave it to work while you brush your teeth or have a wash, and then swish the toilet pan with a toilet brush. If you do this, you should never need to do a heavy-duty clean of your toilet again.

One of the most unhygienic areas in your bathroom can be the floor around the toilet. This can be particularly hazardous if you have young children that get distracted easily, don't get to the toilet in time or would rather be playing than attending to a call from Mother Nature! If you have a hard floor, choose any of the liquid spray cleaners from the Antibacterial cleaners section (see pages 48–50) and keep it in the bathroom. I keep a spray bottle in the bathroom with water and 10 drops each of lavender and lemon essential oils. It smells so uplifting for one of my least favourite jobs! Anytime a little visitor has been to the toilet, give the floor a quick spray and wipe down with a soft cloth. If your bathroom is carpeted, the use of specially shaped rugs that fit around the toilet bowl can be useful. These can be put in the washing machine for cleaning.

Homemade recipes
Seat freshener

It is the seat and under the rim of the seat that are most likely to be unhygienic. The bowl itself is regularly flushed through with water, so germs don't lurk as much in there. Use this bacteria-busting formula to clean around and under the seat and the toilet flush handle.

Place 60 ml (2 fl oz) of liquid castile soap, 450 ml (16 fl oz) of water, 5 drops each of eucalyptus, lemon and tea tree essential oils in a spray bottle and shake well. Keep this spray next to the toilet and use to spray around the toilet seat and handle after use. Wipe away with a soft cloth.

Toilet rim disinfectant
Moving down into the toilet, the rim can become caked with limescale which can breed germs. This simple recipe will work to break down limescale and leave your toilet smelling clean.

Place 115 ml (4 fl oz) of white vinegar, 115 ml (4 fl oz) of water, 4 drops each of lemon and tea tree essential oils in a spray bottle and shake well. Spray this mix around the rim of the toilet and leave to work for half an hour. Scrub with a toilet brush and flush.

Toilet bowl cleaner
Instead of pouring bleach or harsh chemical cleaners down your toilet, you can make something that is safer and doesn't bring tears to your eyes with a strong smell!

Place 60 ml (2 fl oz) of liquid castile soap, 5 drops each of peppermint and lemon essential oils and 30 ml (1 fl oz) water into a squirty bottle and shake. Add 1 tablespoon of white vinegar and mix well. Squeeze this sweet smelling mixture into the toilet bowl and leave to work for an hour. Scrub with a toilet brush and flush.

Soda crystals soak
This formula will keep your toilet bowl clean and prevent blockages. Put on rubber gloves and pour a handful of soda crystals down the toilet pan before bed. Leave overnight. In the morning, scrub with a toilet brush and flush.

Borax deodorizer
Put on rubber gloves and sprinkle 70 g (2½ oz) of borax into the toilet bowl. Swish this around with the toilet brush. Leave to stand overnight for a fresh, clean toilet bowl in the morning.

Toilet bowl fizzer
Pour 250 ml (9 fl oz) of white vinegar into the toilet bowl and scent this with 5 drops of your chosen essential oil if you wish. Add 250 g (9 oz) of baking soda to the toilet bowl. Leave overnight. In the morning, scrub the toilet with a toilet brush before flushing.

The bedroom and sitting room

You probably spend a lot of your leisure time in the sitting room and around a third of your life sleeping in the bedroom! Babies and young children spend a lot of time playing on the floor and even longer sleeping, so it's important to keep both areas as free from pollution and harmful chemicals as possible. Carpet cleaners, and furniture and metal polishes can all contain toxic chemicals. In this chapter, you will find recipes for safer, more natural and effective alternatives.

Dust and dust mites

What is dust?
Dust is a general name for minute particles with diameters smaller than 500 micrometers. Dust is made up of three main components: dead skin cells from humans and pets, the dried faeces and desiccated corpses of dust mites, and other particles, such as clothing fibres and dirt we bring in from the outside.

What are dust mites?
Dust mites are living creatures that are not visible to the naked eye. They feed off dead skin cells found in dust and live in bedding, carpets and soft furnishings. In a ten-week life span, a dust mite will produce approximately 2,000 faecal particles and an even larger number of partially digested enzyme-infested dust particles. A person sheds about 1.5 g of skin cells and flakes every day (around 550 g / 1 lb 3 oz per year), which is enough to feed roughly a million dust mites.

What are dust mite cleaners and what's in them?
These are sprays, gels, powders and liquids that kill the house dust mite. These products contain various chemical pesticides along with preservatives and artificial fragrances.

Dangers to you
You might think of dust as a benign substance, but research at Exeter University's Greenpeace laboratory uncovered 35 hazardous chemicals in house dust, including flame retardants, solvents and petrol additives. Some of these were known to be either carcinogenic, toxic or damaging to the reproductive function and the immune system.

Allergies: dust mites are the most common cause of asthma and allergic symptoms. Because dust mites reproduce quickly, their effect on human health can be significant. Obvious signs of dust mite allergies include itchiness, sneezing, inflamed/infected eczema, watering eyes and a runny nose. It is the enzymes they produce to digest dust particles that provoke allergic reactions.

Pesticides: pesticides kill pests. Many of the synthetic chemicals used to do this are not without their side effects. One off-the-shelf product contains iodopropynyl butylcarbamate, which is a suspected neurotoxin (poisonous to the nervous system) and also toxic to the liver, linked to reduced fertility and more worryingly, reduced chance for a healthy, full-term pregnancy. It has recently been classed as a contact allergen.

Dangers to the environment
Many pesticides are toxic to other species such as bees, fish, aquatic invertebrates and amphibians. If populations of some of these decline, it can have disastrous effects on the eco system.

Keeping it eco-friendly
There are several steps you can take to moderate the amount of dust that's in your home. The following tips will help keep dust under control!

Reducing dust: it's clear that using toxic pesticides is not the answer, but dust and dust mites can be a health concern for some, and so need to be dealt with. As part of the 'prevention is better than cure' routine, don't let dust build up in your home.

Remove shoes: one of the most significant steps you can take is to adopt a 'no shoes' rule in your home. Provide good quality welcome mats at the doors, and get everyone to wipe their shoes and leave them at the door. The majority of contaminants in house dust comes from pollutants brought in from the outside on shoes and clothing.

Regular dusting: you'll notice that electronic equipment such as TVs and computers attract dust rapidly. Once a week or so, have a wipe around with a wrung out damp cloth to prevent dust gathering. For flat-screen computers, use an antistatic cloth and do not use any other products on them otherwise you may damage them. If you have old cotton T-shirts that are beyond wear, cut them up and use them as dusting cloths.

Regular vacuuming: can blow dust back into the room. When you next replace your vacuum cleaner, buy one with a well-sealed, high-quality HEPA filter, which removes almost all airborne particles.

Reducing house dust mites: dust mites thrive in humid, warm conditions. Reducing dampness in your home is a surefire way of diminishing the number of dust mites: leave a door or window open when cooking and keep a window open or an extractor fan running when taking a bath or shower. If a member of your household is showing signs of dust mite allergy, then invest in protective mattress and pillow covers. If they really suffer, you may need to take up carpets in their bedroom and have floorboards instead. Dust mites die when exposed to direct sun rays, so if dust mites are a problem in your home, airing soft furnishings on a hot, sunny day can help.

Top tips for killing dust mites

Hot wash: wash bedding at 60 degrees. This isn't exactly eco-friendly, but it is better than resorting to harsh, toxic chemicals. Dust mites cannot survive at this high temperature.

Deep freeze: freezing will kill the house dust mite. Every other month, place children's soft toys or pillows in the freezer overnight.

Steam clean: steam kills dust mites, so use a steam cleaner on carpets and mattresses every six months. Follow the manufacturer's guidelines for chemical-free cleaning.

Damp dusting: dusting with a dry cloth moves dust from one area in the room to another. Always damp dust, but if you have any antique furniture, use a homemade polish and a soft cloth instead. One of the best solutions to damp dust with is black tea! Make up 600 ml (1 pt) of strong black tea, leave it to cool and use a cloth wrung out in the tea to dust all your surfaces. House dust mites do not like the tannin which is present in tea.

Furniture polishes and waxes

What are furniture polishes and waxes?

Furniture polishes and waxes are highly fragranced sprays, liquids, solid waxes and wipes. They remove dust, clean surfaces and add a protective smear-free shine. They are meant, predominantly, for wooden surfaces, but most can be used on glass, laminated surfaces, sitting room and bedroom furniture and electronics. These products should never be used on floors or bathtubs as they can make the surfaces dangerously slippery.

How do furniture polishes and waxes work?

Many furniture polishes contain antistatic agents to leave surfaces dust-free and shiny. By removing dust, surfaces look cleaner and some polishes add a silicone wax layer to items to produce a sparkling finish. By the use of fragrances, they scent your home too. Most products are sprayed directly onto furniture, about 15 cm (6 in) from the surface, and are wiped immediately with a cloth. Wood cleaners remove ground-in stains from wooden floors, kitchen units and doors – areas that are more likely to get dirty than dusty.

What's in furniture polishes and waxes?

Wood cleaners contain soap, preservatives and surfactants. Surfactants are surface-active agents that reduce the surface tension of a liquid in which they are dissolved, so that stains are suspended and can be removed so they don't go back onto the surface you have just cleaned. Furniture polishes and waxes contain surfactants, strong artificial fragrances and solvents. Some wipes contain iodopropynyl butylcarbamate.

Dangers to you

Most aerosol furniture polishes carry the warning that deliberately inhaling the product may actually kill. It is hard to avoid exposure to furniture polishes because of the nature of the products. You can inhale their toxic chemicals, absorb them through your hands when using them, particles travel through the air into other parts of your home and most leave an intentional residue on surfaces to leave them shiny.

Phenol, also known as carbolic acid, has been used in furniture polish throughout history. It is an effective antiseptic, but is also a dangerous substance if inhaled, ingested or if it comes into direct contact with the skin. Inhalation can cause headaches, dizziness, fatigue and in some rare cases, can be fatal. Skin contact with phenol can cause burns, hives and rashes. Nitrobenzene, another chemical commonly used in polishes, is highly toxic, readily absorbed through the skin, and can cause severe damage to the central nervous system, so always wear rubber gloves when using furniture polish.

Dangers to the environment
Propellants used in most aerosol cans are highly flammable and the base waxes in solid polishes are often derived from petroleum, which is a non-renewable resource. As most furniture care products come in the form of aerosol sprays, the chemical ingredients can travel in the air and pollute the environment. In addition, propellant gases used in aerosols to force the product out of the can, contribute to global warming.

Keeping it eco-friendly
If you think about the strength of a tree, you'll understand that wooden furniture is pretty robust! How you care for your furniture is more important than spraying it with polishes and using waxes. As long as you keep wet away from wooden furniture by using drinks coasters and mopping up spills straight away, then wood will virtually take care of itself. Don't allow children or pets to gouge or scratch the wood. To keep your wooden furniture in peak condition, refer to the recipes opposite. Furniture sprays and polishes containing silicone can eventually build up a haze, so it is best to avoid these products. In addition, many polishes and waxes contain water that can eventually lead to spotting on wooden surfaces. It seems that these products can actually do more harm than good!

Top tips for cleaning wood

White vinegar: makes a great cleaner for any type of wood, including wooden floors. Spray it on neat and wipe with a soft cloth.

Olive oil: straight from the kitchen cupboard, olive oil will add a beautiful shine to your furniture and help protect it. Apply a few drops with a soft cloth and buff.

Olive oil and white vinegar wood 'food': combine 3 parts olive oil with one part white vinegar in a bottle and shake to mix. This will create a nourishing product that wood will drink up. Add 10 drops of your favourite essential oil to the mix, if you wish. Apply this mixture onto wooden furniture with a cloth, rub in and then buff with a separate dry cloth. Some people prefer to use a higher ratio of vinegar to oil and feel this works better. Why not try both and decide for yourself? Combine 60 ml (2 fl oz) of white vinegar with a few drops of jojoba or olive oil, and 2 drops of lemon essential oil. Apply a small amount with a soft cloth to your furniture and buff with a separate clean, dry cloth.

Scented beeswax

Many 'premium' furniture waxes contain beeswax, so why not make your own? This simple recipe will leave your furniture looking clean and beautifully nourished. In a heatproof glass bowl over a saucepan of simmering water, place 3 tablespoons each of beeswax and either olive, walnut or jojoba oil along with 1 tablespoon of white vinegar. Stir gently until the ingredients have melted, then remove the saucepan from the heat and leave to cool for 10 minutes, keeping the bowl over the pan. Stir in 5 drops of your favourite essential oils (I like 3 of lemon and 2 of lavender). When the bowl is cool enough to handle, pour the mix into a glass jar with a lid. Use the wax as you would any regular wax cleaning polish, with a soft cloth. Buff after applying.

Natural beeswax is available from your local beekeeper, on the Internet and from some herbal stores.

Metal polishes

What are metal polishes?
Metal polishes are liquids or wipes that clean most metal surfaces, including brass, copper, silver and stainless steel. They degrease as well as remove dust, dirt, tarnish and fingerprints, leaving your surfaces bright and shiny.

How do metal polishes work?
Most metal polishes contain strong acids to dissolve alkaline tarnishes. The addition of perfumes leaves your surfaces and items smelling clean and fresh.

What's in metal polishes?
Strong acids such as phosphoric, sulfuric or hydrofluoric acids neutralize alkaline tarnishes. Other ingredients are caustics, such as ammonia. In addition, you'll find surfactants, preservatives and perfumes.

Dangers to you
Even in diluted forms, acids can burn and scar the skin. They are very dangerous to the eyes and can cause irreversible damage. Hydrofluoric acid is corrosive and a contact poison. It can have devastating effects if allowed to come in to contact with the skin. It can be fatal if as little as 2.5 per cent of total body surface area is exposed to concentrated hydrofluoric acid.

Ammonia is irritating to the eyes and mucous membranes. Mixing it with chlorine-containing products or strong oxidants such as household bleach, can create hazardous compounds such as chloramines.

One particular silver cleaner contains thiourea. It carries the warning on the label 'Possible risk of irreversible effects'. Hepatic tumours have resulted from chronic administration of thiourea in animal tests. Thiourea can also enlarge the thyroid gland.

Dangers to the environment
Ammonia, even at diluted concentrations, is highly toxic to aquatic animals, and for this reason is classified as dangerous for the environment.

Keeping it eco-friendly
Many metal polishes contain so many toxic ingredients that prevention is better than cure. Unfortunately, there is no magic way to prevent tarnish from forming, but after you have cleaned a metal product using one of the recipe below, a couple of drops of olive or jojoba oil rubbed in can help slow down the tarnishing process.

Homemade recipes

Brass and copper: this recipe for cleaning brass and copper is so easy and works really well. Cut a lemon or lime in half, sprinkle salt over the cut surface and use it like a scourer. It's perfect on copper saucepans.

Silver: silver tarnish can be removed from solid silver objects with aluminium foil, salt and baking soda. Before you start, open a window (as the tarnish is removed, it can release a foul smelling gas). Line the bottom of a glass or plastic bowl, or washing up bowl with aluminium foil. Three-quarter fill the container with warm water and dissolve 1 tablespoon of salt and 1 tablespoon of baking soda in the water. Place the silver items in the bowl and leave to soak for 15 minutes. Remove the items from the bowl and wipe down with a soft cloth.

Stainless steel: stainless steel can be effortlessly cleaned with one safe ingredient: baking soda. Sprinkle some baking soda onto a damp soft cloth or sponge and use to gently bring shine to stainless steel surfaces. Rinse with warm water and follow with a 50/50 mix of white vinegar and water.

Carpet cleaners

What are carpet cleaners?
Products for carpets include deodorizers, shampoos and spot removers. Carpet deodorizers usually take the form of a powder which you sprinkle onto your carpet, leave to work and then brush or vacuum away. They are designed to 'freshen up' your carpet by covering up smells and are aimed at pet owners, smokers or those with young children. Carpet shampoos are concentrated detergents, designed to absorb grease and oil. Spot removers contain ingredients that digest and bleach stains.

What's in carpet cleaners?
Most conventional carpet cleaning products contain a cocktail of artificial fragrances, optical brighteners, pesticides, fungicides and solvents.

Dangers to you
There are many different chemicals in the wide variety of carpet cleaners available. Many carpet shampoos contain optical brighteners, just like laundry products, which give the appearance of a brighter, cleaner surface. Optical brighteners remain on the carpet, giving the illusion of cleanliness, but they can cause skin irritation.

Spot cleaners are the most toxic of all carpet cleaning products. Most spot removers contain perchlorethylene and naphthalene, both of which are poisonous if swallowed. Perchlorethylene is a solvent which is highly irritating and has a lingering smell. It has been shown to damage the central nervous system and contribute to kidney and liver damage. Short-term effects of exposure to perchlorethylene include dizziness, light headedness, headaches and disorientation. Naphthalene is a pesticide, insecticide and fungicide. It is a neurotoxin that irritates the eyes, skin and lungs. Naphthalene is very toxic to children.

Dangers to the environment
Carpet cleaning products often contain phosphates, which can result in excessive algae growth once you have poured it into the water course. Warnings on one product read 'Extremely flammable'. Without adequate ventilation, formation of explosive mixtures may be possible.

Keeping it eco-friendly

If your carpets are smelly, there is a reason for this. Refer to the Air fresheners section for hints (see pages 40–41). It's not always easy to keep wall to wall carpets clean. We can't take them outside and air them, or get to the backs of them to give them a thorough clean. Every time you walk on a carpet, you send up a cloud of dust and dirt. The Institute for Total Carpet Hygiene (ITCH) states 'If you haven't washed your carpet in the last 12 months you can be sure that lurking deep in the carpet fibres will be bacteria, fungus, chemicals, dead skin, dust, food particles, germs, pollen, grease, grit and dust mites. If neglected, these particles will work their way deeper into the carpet and may contribute to health problems including allergies, asthma and emphysema'.

Six steps to clean carpets

Prevention is the key to all successful green household cleaning. Follow these tips for clean, healthy carpets:

1. Shoes off! Make your home a shoe-free zone. Place good quality welcome mats at the entrance to your home. Ask guests to wipe their shoes and leave them at the door.

2. Vacuum regularly so that dirt and dust don't accumulate. Vacuum at least once a week and consider using a carpet sweeper in between times to gather any fluff and dirt.

3. Mop up! Deal with spills as they happen. Remember to blot, not rub stains, otherwise you push them further into the carpet. As soon as a spill touches the carpet, pour soda water over it to fizz up the stain. Then immediately put some old clean towels over the area and stamp on them to draw out as much liquid as possible. Keep going until the carpet is just damp.

4. Steam clean your carpets at least once a year to keep them clean and the fibres in good condition.

5. Use rugs or mats over your carpets. Rugs can be taken outside and aired, and some can be washed in a washing machine. These help to protect your carpets and keep them clean.

6. If your carpet has seen better days, why not switch to hard floors and rugs? Hard floors should be brushed once a day.

Homemade carpet deodorizers

One of the simplest and safest carpet deodorizers is baking soda.
Add dried, crushed lavender flowers if you wish, or a few drops
of your favourite essential oil.

Baking soda scented deodorizer

In a lidded, wide-topped jar, mix together 1 tablespoon
of dried, crushed herbs of your choice with 250 g (9 oz)
of baking soda. Sprinkle liberally around a freshly-vacuumed
or brushed carpet and leave to work overnight. The following
day, brush up any excess baking soda mix with a clean dustpan
and brush. Vacuum thoroughly to remove any last traces of the mix.

To personalize your scent, select the herbs to match your mood. Dried
peppermint will give an invigorating, fresh smell. Lemon balm is relaxing.
Rose petals are warming and floral. If you don't have any dried herbs, baking
soda alone will work wonders. It won't add a fresh smell, but it will effectively
remove and neutralize offensive odours.

Homemade carpet shampoos

One of the best ways to really clean your carpets is with a steam cleaner.
You don't need to use a shampoo if you steam clean as the heat will lift
dirt and kill dust mites, bacteria and fleas in the process.

Shampoo recipe used with carpet cleaning machines

If you want to shampoo a grubby carpet, then it is easy to make your own.
The following mix can be put into the cleaning reservoir of a carpet cleaner.
Mix 4–5 teaspoons of liquid castile soap or non-toxic dishwashing liquid in
9 litres (2 gallons) of hot water. For really dirty carpets, add ½ teaspoon of
borax or soda crystals per litre (1¾ pt) of water.

Shampoo recipe for hand cleaning carpets

This recipe can be made in a large bowl or bucket if you want to clean
your carpet by hand. Combine 1 part of liquid castile soap or non-toxic
dishwashing liquid with 1 part of hand-hot water. Work the suds into the
carpet with a brush and blot dry with old towels. The carpet should be damp,

not wet. Allow to dry thoroughly before vacuuming. If you saturate the carpet, you risk getting damp in the backing, which could potentially lead to mildew and carpet shrinkage, so go lightly!

Spot cleaners
Borax is an effective spot remover. Make sure you test on an inconspicuous area of carpet first. Although borax is safer than many conventional products, it is still toxic, so wear a pair of rubber gloves.

Dissolve 125 g (4½ oz) of borax in 1 litre (1¾ pt) of hot water. Apply with a stiff brush and work from the outside of the stain inwards. Remember not to saturate the carpet, just keep it damp. When you have finished, absorb any excess moisture with old towels.

Specific stain removal
Tackle stains as soon as they appear for maximum chance of removal. Apart from wax, never treat a stain with heat as this can set the stain.

Blood: wet the stain with cold water and keep blotting dry with old towels. The blood should soon lift out. Another trick is to use soda water. Soda water fizzes the blood out, which you can then blot with old towels. An old fashioned way is to use spit, as it contains natural enzymes, especially if the spit belongs to the person whose blood it is. Spit on the stain and blot it with old cloths. Obviously you can only do this on a few spots of blood.

Chocolate, ketchup, tea or coffee: scrape as much of the chocolate or ketchup away as you can (or soak up as much tea or coffee as possible), then pour soda water over the stain to fizz it out. Keep reapplying soda water and blotting dry with old towels. If this doesn't work, try white vinegar; test out on an inconspicuous area of carpet first. Spray the stain with a 50/50 mix of cold water and white vinegar, and keep blotting with old towels.

Lipstick or bubblegum: put 5 drops of pure eucalyptus essential oil onto a cloth and blot at the stain. If you are working at bubblegum, blot the stain and then scrape away the bubblegum with a knife. Test on an inconspicuous area of carpet first. If there is no discolouration, you can proceed.

Mud: leave the mud to dry and then brush up with a stiff brush. It should lift away easily. If a stain remains, then sprinkle with baking soda, leave to work for a few hours, and brush away. Although it may not seem right to leave things to dry on the carpet, trying to deal with mud when it's wet can lead to a worse stain that is more difficult to remove!

Red wine: probably one of the most stubborn stains to get rid of! Pour soda water onto the stain immediately and blot with old towels. Repeat until the stain fizzes out. If you're lucky, this is all you'll need to do. If some stain remains, apply salt and leave to work for 10 minutes, then brush away the salt. Failing that, apply borax spot remover (see page 121) with a soft sponge. Leave for 30 minutes and repeat if necessary. If the stain persists, try 3 per cent hydrogen peroxide on an inconspicuous area of your carpet first. If there is no discolouration, then dab the stain gently with the hydrogen peroxide.

Urine: pets tend to return to the same place in the house if they pee on the carpet so you'll need to neutralize the odour (and train your pet!). The simplest way of removing urine stains is to use baking soda. Blot the excess urine with old towels. This can take quite a while and a lot of towels, but is very effective. When the carpet is just damp, sprinkle the area liberally with baking soda and leave overnight. The following day, brush up any excess baking soda and vacuum the carpet. If the smell is still there, use a 50/50 mix of white vinegar and cold water to spray on the stain (test on an inconspicuous area of your carpet first). Make the carpet damp again with the vinegar mix, blot up any excess with old towels and air the room.

Wax: leave the wax to dry and then scrape as much as you can away with a knife. To remove the remaining bits of wax, place a clean cloth, paper bag or blotting paper over the wax and iron over it with a warm, but not hot iron; the wax will melt onto the cloth or bag. Another option is to apply crushed ice cubes directly onto the wax. This makes it hard and brittle so that you can pick it off.

Floor cleaners

What are floor cleaners?
Floor cleaners are liquids, wipes, sprays and creams that clean most flooring such as vinyl, sealed laminate, linoleum and glazed tiles.

How do floor cleaners work?
Floor cleaners contain ingredients that cut through grease and grime and most contain antibacterial agents. They are quick-drying and highly fragranced to leave your home smelling clean. Liquids are usually diluted in a bucket of warm water and used with a mop. Creams are applied with a sponge and sprays are applied and wiped with a cloth. Impregnated wipes are used once and disposed of.

What's in floor cleaners?
Most conventional floor cleaners contain artificial fragrances, disinfectants, surfactants, solvents and preservatives.

Dangers to you
Ethanol is a widely used solvent which allows for the quick-drying of products. It can make up to 15 per cent of the product in floor wipes and is a central nervous system depressant. Floor cleaning products should not come in to contact with the skin or eyes as they can cause irritation. Minute particles of toxic chemicals in spray products can be inhaled easily or travel to less desirable areas in your home, such as food preparation surfaces or dining tables. Remember that most babies and young children will crawl or play on the floor and pick up items from the floor to put into their mouths.

Keeping it eco-friendly
A quick sweep with a brush and wiping away dirty marks will mean you only need to do a thorough clean once a week. As soon as you spill anything on a hard floor clear it up. Use a damp cloth, then dry the floor. Place good-quality welcome mats at the entrance to your home and adopt a no-shoes rule in

the house. If you have dogs that leave muddy footprints, get them to stand on a towel when they come into the house while you clean their feet and legs as best you can. If your hard floors look dull rather than shiny, then there is probably build-up of detergent from conventional products. Spray a 50/50 mix of white vinegar and cold or warm water onto the floor, and polish with a cloth to remove the residue and restore shine.

Homemade recipes for hard floors

White vinegar: white vinegar is an excellent floor cleaner that cuts through grease and grime. It can be used on virtually any hard floor type such as vinyl, linoleum, tile and wood. Add 225 ml (8 fl oz) of white vinegar to a bucket of hot water. Add 5 drops of your favourite essential oil if you wish. Use with a damp mop to clean your floors.

Soap: add 2 tablespoons of liquid castile soap to a bucket of hot water. Use castile that is already scented with oils such as peppermint, or add a few drops of your own. Use with a damp mop to clean your floors.

Detergent: a good quality, eco-friendly, dishwashing liquid can work as well as a specialized floor cleaning product. Add a squirt of dishwashing liquid to a bucket of hot water. Use this solution with a damp mop or cloth to clean your floors. The secret with this recipe is not to use too much detergent or you will be left with lots of foam that attracts dirt. Follow this with a 50/50 mix of white vinegar and cold water to remove any last traces of detergent if you need to.

Soda crystals: a strong solution of soda crystals is perfect if your floor is really greasy or has ground-in dirt. Exercise caution though, as soda crystals are powerful enough to remove any wax or sealant on the surface! Wear rubber gloves and dissolve 250 g (9 oz) of soda crystals in 600 ml (1 pt) of hot water. Use a cloth dipped in this solution to wipe over areas of your floor.

Water: it's good to remember how effective hot water can be on its own. Use a bucket of hot water with 5 drops of your chosen essential oils added to freshen up a floor or remove minor spills. Another idea is to fill an average size 500 ml (18 fl oz) spray bottle with cold or warm water and add 10 drops of your favourite essential oils. If you spray this mix onto your floors and use an eMop on a daily basis, you will never need to scrub at your floors. Spending 5 minutes a day giving your floors a quick wipe over will save you time in the long run!

Natural air filters

Energy-efficient homes

In recent years, our homes have become more energy efficient thanks to double glazing and insulation. This is great for fuel bills, but one of the reasons why indoor air pollution can be higher than outdoor. Any chemicals released into the air from household cleaners, personal care products, soft furnishings and carpets stay in your home and do not dissipate into the open air.

Tackling indoor air pollution

If you are still worried about indoor air pollution despite making your own cleaning products, then there is some good news. There is a way to purify air that will bring beauty and colour into your home. Welcome to the amazing world of houseplants! According to extensive research by NASA, many common houseplants help fight indoor air pollution.

Many common houseplants absorb benzene, formaldehyde and trichloroethylene, which are three of the worst offenders found around the home in soft furnishings, fixings, MDF, computers and TVs, as well as household cleaners.

Using plants to clean the air

Following NASA research, the recommendation is that you use 15 to 18 good-sized houseplants in 15–20 cm (6–8 in) diameter containers to improve air quality in an average 170 square metre (1,800 sq ft) house.

To remove trichloroethylene: the best plants for removing trichloroethylene are Gerbera Daisy (*Gerbera janesonll*), Marginata (*Dracaena marginata*) and Peace Lily (*Spethiphylium 'Muana Loa'*).
To remove benzene: the best plants for removing benzene are Gerbera Daisy (*Gerbera janesonll*), Pot Mum (*Chrysanthemum morifollum*) and Peace Lily (*Spethiphylium 'Muana Loa'*).
To remove formaldehyde: the best plants for removing formaldehyde are Bamboo Palm (*Chamaedorea selirizll*), Janet Craig (*Dracaena deremensis 'Janet Craig'*) and Mother-In-Law's tongue (*Sansevieria laurentii*).

Books

Ashton, Karen – *The toxic consumer: How to reduce your exposure to everyday toxic chemicals* (Impact Publishing Ltd, 2007)

Grace, Janey Lee – *Imperfectly natural home: The organic bible* (Orion, 2008)

Logan, Karen – *Clean house, clean planet: Clean your house for pennies a day, the safe, nontoxic way* (Simon & Schuster, 1997)

Martin, Angela – *Natural stain remover: Clean your home without harmful chemicals* (Apple Press, 2003)

Thomas, Patricia – *What's in this stuff?: The hidden toxins in everyday products and what you can do about them* (Perigee Books, 2008)

Wolverton, B.C. – *How to grow fresh air: 50 houseplants that purify your home or office* (Weidenfeld & Nicolson, 2008)

Websites

More information on green cleaning and the author:
http://selfsufficiencyhouseholdcleaning.com

Dri-pak: www.dri-pak.co.uk
(mail order borax, baking soda, soda crystals and vinegar)

Flylady: www.flylady.net
(get your life in order and your home sparkling clean)

Acknowledgements

Rachelle would like to thank: Richard, my soul mate, for your unconditional love. Verona for your continual inspiration. Detta for your unwavering love and Happy Dances. Corinne for your tireless work and enthusiasm. Sue for being my earth angel. Veronika for believing in me from the start. I feel blessed to have you all in my life. Peggy, whose smile lights up my life.